Medical Device Regulations: Transitioning from MDD 93/42/ EEC to MDR 2017/745

Medical Device Regulations: Transitioning from MDD 93/42/EEC to MDR 2017/745

Shalinee Naidoo

ARCLER PRESS

www.arclerpress.com

Medical Device Regulations: Transitioning from MDD 93/42/EEC to MDR 2017/745

Shalinee Naidoo

Arcler Press
224 Shoreacres Road
Burlington, ON L7L 2H2
Canada
www.arclerpress.com
Email: orders@arclereducation.com

ISBN: 978-1-77407-808-2 (Hardcover)

ABOUT THE AUTHOR

Shalinee Naidoo is currently the Regulatory and Product Development Manager of a medical device manufacturer based in South Africa. She is directly involved in global regulatory compliance and design and development of new medical devices from idea conception to market. She is also the founder of Life of Shal (www.lifeofshal.com) an online travel journal created to inspire others to explore the world and Scientist's Sanctuary (www.scientistssanctuary.com) – a science communication company that specializes in bridging the gap between scientific knowledge and creative communication for both the academic and corporate world.

TABLE OF CONTENTS

GLOSSARY

A

Active implantable device – Any medical device that relies on a source of electric energy or any other source of power other than that directly generated by the human body or gravity which is intended to be totally or partially introduced, surgically or medically, into the human body or by medical intervention into a natural orifice, and which is intended to remain after the procedure.

Adverse event – Any event or problem that can or does result in permanent impairment, injury, or death to the patient or the user of a medical device.

Audit – A systematic, independent examination of a Manufacturer's quality system that is performed at defined intervals to determine: (1) whether both quality system activities and the results of such activities comply with quality system procedures, (2) that these procedures are implemented effectively, and (3) that these procedures are suitable to achieve the set out quality system objectives.

B

Body orifice – Any natural opening in the body. This includes the external surface of the eyeball or any permanent artificial opening.

C

CE mark – symbol required for all medicines marketed into the European Union indicating that they meet common standards of performance and safety, known as essential requirements (ER).

Clinical data – Safety and/or performance information that are generated from the clinical use of a medical device.

Clinical evaluation – The assessment and analysis of clinical data related to a medical device in order to verify the clinical safety and performance of the device when used as intended by the manufacturer.

Clinical evidence – The clinical data and the clinical evaluation report pertaining to a medical device.

Clinical evidence for an IVD medical device – All the information/data that supports the scientific validity and performance for its use as intended by the manufacturer.

Clinical investigation – Any systematic investigation or study in or on one or more human subjects, undertaken to assess the safety and/or performance of a medical device. This term is similar in meaning to 'clinical trial' and 'clinical study.'

Clinical trial – A clinical investigation or research study involving one or more human subjects which generates clinical data to be used to demonstrate the safety and effectiveness of a medical device.

Conformity assessment – Systematic examination of evidence generated and procedures undertaken by a manufacturer, under requirements established by the Regulatory Authority. This is done to determine that a medical device is safe and performs as intended by the manufacturer and aligns or 'conforms' to the Essential Principles (EP) of Safety and Performance for Medical Devices.

Conformity assessment body (CAB) – A body, other than a Regulatory Authority, engaged in determining whether the relevant requirements in technical regulations or standards are met.

Contract sterilizer – A person or business that provides sterilization services for another establishment's devices.

D

Distributor – May also include multiple distributors of the same device. This is a person or business that receives finished devices from another establishment for the purpose of offering them for commercial sale.

E

Effectiveness – A device is clinically effective when it produces the effect intended by the manufacturer relative to the medical conditions it is meant to treat.

H

Harm – Physical injury or damage to the health of people or damage to property or the environment.

Hazard – Potential source of harm.

I

Immediate danger – A situation where the patient is at risk of either losing their life or an important physiological function if no immediate preventative measure is taken.

Implantable device – Any device, including those that are partially or wholly absorbed and which are intended to be partially introduced into the human body through surgical intervention and intended to remain in place after the procedure for at least 30 days is also considered an implantable device.

Importer – A person or business that receives devices from foreign manufacturers for the purpose of offering them for commercial sale and/or distribution in the recipient country.

***In vitro* diagnostic (IVD) medical device** – A medical device, whether used alone or in combination, intended by the manufacturer for the *in vitro* examination of specimens derived from the human body solely or principally to provide information for diagnostic, monitoring or compatibility purposes.

Incident – An unusual or unexpected event that is associated with the use of a medical device and which may or may not lead to problems.

Indications for use – A general description of the disease or condition the medical device or IVD medical device will diagnose, treat, prevent, cure, or mitigate, including a description of the patient population for which the medical device or IVD medical device is intended.

Instructions for use – Information provided by the manufacturer to inform the user of the medical device of the intended purpose and proper use and of any precautions to be taken. This is also known as a "package insert."

Intended use/purpose – The objective or intent of the manufacturer regarding the use of a product, process or service as reflected in the specifications, instructions, and information provided by the manufacturer.

International medical device regulators forum (IMDRF) – A voluntary group of medical device regulators from around the world who have come together to build on the strong foundational work of the GHTF (global harmonization task force). This group aims to accelerate international medical device regulatory harmonization and convergence.

Invasive device – A device which, in whole or in part, penetrates inside the body, either through a body orifice or through the surface of the body.

J

Jurisdiction – The power, right, or authority to interpret and apply regulatory laws.

L

Label – Written, printed, or graphic information either appearing on the medical device itself, or on the packaging of each unit, or on the packaging of multiple devices.

Labeling – The label, instructions for use, and any other information that is related to identification, technical description, intended purpose, and proper use of the medical device, but excluding shipping documents.

Lay user – An individual who does not have formal training in a relevant field or discipline.

Life cycle – All phases in the life of a medical device, from the initial conception to final decommissioning and disposal.

Life supporting or life-sustaining – A device that is essential to, or that yields information that is essential to, the restoration or continuation of a bodily function important to the continuation of human life.

Local manufacturer – A manufacturer within the country which the medical device is being sold.

Long term – Normally intended for continuous use for more than 30 days.

M

Manufacturer – A person or business that produces an article that meets the legal definition of a medical device.

Market surveillance activities carried out and measures taken by competent authorities (CA) to check and ensure that medical devices comply with the requirements of Regulations set out.

Medical device – Any instrument, apparatus, implement, machine, appliance, implant, reagent for *in vitro* use, software, material or other similar or related article intended by the manufacturer to be used, alone or in combination for human beings, for one or more medical purposes.

Medical products – Medicines, vaccines, diagnostics, and medical devices.

P

Patient – An individual under the care of a healthcare provider who may benefit from the action of a medical device.

Performance – The ability of a medical device to achieve its intended purpose as stated by the manufacturer.

Placing on-market control – All activities that performed by local establishment (Importer/ distributor/Authorized Representative, etc.) such that it complies with regulatory requirements to ensure safety and effectiveness of medical devices.

Post-market surveillance – Carried out systematically and proactively by the manufacturer to collect and review information/experience gained from their medical devices such that they ensure continued safety and efficacy.

Post-market – When medical devices/products are already on the market.

Post-market control – All activities that are performed by all parties (either manufacturers or local establishments) such that they comply with regulatory requirements to ensure continued safety and effectiveness of medical devices during their use.

Pre-market control – All activities that are performed on a medical device by a manufacturer, either local or international, such that it complies with regulatory requirements to ensure safety and effectiveness.

Problem – Possible faults with the medical device, difficulties in using the device or any undesirable outcome associated with the use of the device. A problem may not lead to an adverse event but corrective or preventive actions are required.

Q

Quality management system – The organizational structure, responsibilities, procedures, processes, and resources for implementing quality management.

R

Refurbisher/remarketer – A person or business that assumes ownership of previously-used devices and reconditions them for resale.

Regulatory authority (RA) – A government body or other entity that exercises a legal right to control the use or sale of medical devices within its jurisdiction, and that may take action to ensure that medical products marketed within its jurisdiction comply with legal requirements.

Relabeler – A person or business that changes the content of the product labeling offered by the original manufacturer for distribution under its own name.

Repackager – A person or business that packages finished devices from bulk or repackages devices made for shipment in multiple containers.

Reusable surgical instrument – Instrument intended for surgical use by cutting, drilling, sawing, scratching, scraping, clamping, retracting, clipping or other surgical procedures, without connection to any active medical device and which are intended by the manufacturer to be reused after appropriate procedures for cleaning and/or sterilization has been carried out.

Risk – The combination of the probability of occurrence of harm and the severity of that harm.

Risk assessment – Overall process comprising a risk analysis and a risk evaluation.

Risk evaluation – Procedure based on the risk analysis to determine whether the tolerable risk has been exceeded.

S

Safety – Free from unacceptable risk.

Shelf-life: The period of time until the expiry date during which a medical device in its original packaging maintains its stability under the storage conditions specified by the manufacturer.

Short term – Normally intended for continuous use for not more than 30 days.

Stability: The ability of a medical device and IVD medical device to maintain its safety and performance characteristics within the manufacturer's specifications over a specified period of time.

Summary of technical documentation (STED) – A summary of technical documentation held or submitted for conformity assessment purposes.

Surgically invasive device – An invasive device that penetrates inside the body through the surface of the body, with the aid of, or in the context of a surgical operation.

T

Technical documentation – Documented evidence which is normally an output of the quality management system and which demonstrates compliance of a device to the Essential Principles (EP) of Safety and Performance of Medical Devices.

Transient – Normally intended for continuous use for less than 60 minutes.

U

User – The person (professional, lay, or patient) who uses a medical device.

V

Vendor – Any person who sells medical devices, such as a manufacturer, an importer, a distributor, a wholesaler, or a retailer.

Vigilance – Reporting of serious incidents and field safety corrective actions by manufacturers to the relevant competent authorities (CA).

LIST OF FIGURES

LIST OF TABLES

LIST OF ABBREVIATIONS

AIMDD	active implantable medical device directive
AR	authorized representatives
CA	competent authorities
Class A	low risk
Class B	low moderate risk
Class C	moderate high risk
Class D	high risk
EP	essential principles
ER	essential requirements
GHTF	global harmonization task force
GSPR	general safety and performance requirements
ISO	International Standard Organization
IVD	in-vitro diagnostic
IVDMDD	in vitro diagnostic medical device directive 98/79/EC
MDD	medical device directive
MDR	medical device regulation
MEDDEV	medical device document
NBs	notified bodies
QMS	quality management system

PREFACE

Medical devices are important to providing adequate health care and to improving the health of individuals all around the world. Without properly regulated medical devices many routine medical procedures, whether they be bandaging an open wound or implanting an artificial hip, would be impossible. Advancing technology is producing an overwhelming number of new devices onto the market and as such, one of the biggest issues affecting the medical device industry is the diversity of devices available.

In recent times, there has been an unprecedented level of public interest regarding the regulation of medical devices in Europe. This is because devices cannot be placed on the European market without conforming to the strict safety requirements of the European Union. The release of the new medical device regulation (EU MDR 2017/745) in the Official Journal of the European Union in May of 2017 marked the start of a three-year transition period for manufacturers, suppliers, notified bodies (NBs), and national Competent Authorities (CA). Compliance to the new regulation was set to take effect in May of 2020 but has now been postponed by a year due to the coronavirus pandemic. The scale of the changes required to be made is significant and has put a great deal of pressure on manufacturers. This volume aims to provide an easy-to-understand guide for beginners to the medical device regulations (MDRs) in Europe. It looks specifically at the changes between the current Medical Device Directive (MDD) 93/42/EEC and forthcoming Medical Devices Regulation (EU MDR 2017/745) with regards to classification, conformity assessments, and assessing general safety and performance requirements (GSPR).

CHAPTER

1

Introduction

CONTENTS

INTRODUCTION

Medical devices form an integral role in health care delivery and for this reason, ensuring the provision of safe and effective medical products are imperative in today's day and age. These devices cost governments a substantial amount of money and can cost even more if not regulated properly. In 2000, it was estimated that 1.5 million different medical devices were available on the market worldwide. As innovation and rapid advancement in technologies take over, medical devices have now become one of the fastest-growing industries globally. Regulatory controls for medical devices are often complicated to implement due to the many different types of devices available on the market. Often the risks associated with each type of device differ, making the opportunity for misuse considerable. For these reasons, it is not justifiable to subject all types of devices to the highest level of regulatory control and many countries tend to have adopted controls from other countries. In order to sell medical devices within the European Union (EU), medical device manufacturers will need a valid MDD (medical device directive) certificate. Collectively known as the MDD, this core legal framework consists of three directives that regulate the safety and marketing of medical devices in Europe. These directives came into effect in the 1990s and are as follows:

- Active implantable medical device directive (AIMDD 90/385/EE);
- Medical device directive (MDD 93/42/EEC);
- *In vitro* diagnostic medical device directive (IVDMDD 98/79/EC).

New and emerging technologies as well as various well-known incidents within the medical device industry have challenged the current framework, highlighted gaps, and pointed to a certain scarcity of skills and expertise. For this reason, there was increasing need to update the MDD which led to the development and release of the medical device regulation (EU MDR 2017/745) which has essentially put stricter regulations on the entire product life cycle. Under the new EU medical device regulation (MDR) (2017/745), manufacturers will be assessed according to these new regulations and will thus require new certificates for their devices and products. Companies wishing to continue selling their medical devices and products in the EU after enforcement of the MDR will need to transition their technical product documentation as well as the documentation within their quality management system (QMS). Many manufacturers are unaware of how long the transition to the new MDR regulation actually takes and for this reason, ensuring a thorough understanding of what the requirements are asking is imperative for success. If manufacturers do not start planning early enough, it could lead to no valid certificate being issued for a long period of time which in turn may result in businesses not being able to access the European market. Non-compliant notified bodies (NBs), similarly, may lose re-designation. Since the MDR's release in 2017, many regulatory experts have published and circulated resources aimed at helping companies understand the new rules and adjust accordingly. This volume aims to do the same but on a much lower level, offering simple explanations of the major changes in relation to the Directives/Regulations associated with device classification, conformity assessments and general safety and performance requirements (GSPR).

1.1. HOW ARE MEDICAL DEVICES REGULATED IN EUROPE?

Medical devices cannot be placed on the European market without a manufacturer proving conformance to the strict safety requirements of the European Union which, once done, allows for the affixation of the CE conformity mark on the device itself. These safety requirements are laid out in relevant regulations, previously known as the MDD and now the MDR. The MDD formed the core legal framework in Europe since the 1990s. It consists of three directives that regulate the safety and marketing of medical devices. These three directives are:

- Active implantable medical device directive (AIMDD 90/385/EE);
- Medical device directive (MDD 93/42/EEC); and
- *In vitro* diagnostic medical device directive (IVDMDD 98/79/EC).

These have been supplemented since by several necessary updates and relevant guidance documents and directives to guide manufacturers when implementing the regulation. Often these guidance documents and directives are released due to new and emerging technologies which challenge current frameworks or highlight gaps or areas of confusion.

The MDD defines a medical device as:

'Any instrument, apparatus, appliance, material or other article, whether used alone or in combination, including software necessary for its proper application intended by the manufacturer to be used for human beings for the purpose of:

- Diagnosis, prevention, monitoring, treatment or alleviation of disease;
- Diagnosis, monitoring, treatment, alleviation of or compensation for an injury or disability;
- Investigation, replacement or modification of the anatomy or of a physiological process;
- Control of conception and which does not achieve its principal intended action in or on the human body by pharmacological, immunological, or metabolic means, but which may be assisted in its function by such means.'

The CE mark is a conformity mark which all European medical devices must have before they can be placed on the European market. By placing this mark on a device, it is seen as a declaration by the manufacturer that the product meets all the appropriate provisions of the relevant legislation including those related to safety and, where required, has been assessed in accordance with these.

Article 17 (CE marking) of the MDD 93/42/EEC, states:
1. *'Devices, other than devices which are custom-made or intended for clinical investigations, considered to meet the essential requirements (ER) referred to in Article 3, must bear the CE marking of conformity when they are placed on the market.* 2. *The CE marking of conformity, as shown in Annex XII, must appear in a visible, legible, and indelible form on the device or its sterile pack, where practicable and appropriate, and on the instructions for use. Where applicable, the CE marking must also appear on the sales packaging.* *It shall be accompanied by the identification number of the notified body responsible for implementation of the procedures set out in Annexes II, IV, V, and VI.* 3. *It is prohibited to affix marks or inscriptions which are likely to mislead third parties with regard to the meaning or the graphics of the CE marking. Any other mark may be affixed to the device, to the packaging or to the instruction leaflet accompanying the device, provided that the visibility and legibility of the CE marking is not thereby reduced.*

The CE conformity mark consists of the initials 'CE' and is usually represented as in the figure below:

There are specific requirements are the visual display of the CE mark as well, including that the various components of the CE mark must have the same vertical dimension, which may not be less than 5 mm (although this minimum dimension may be waived for small-scale devices). It is also important to note that the CE marks or rather the letter "CE" does not stand for any specific words but rather it is a symbol indicting a declaration by the manufacturer that the product meets all the appropriate provisions of the relevant legislation. It should not be placed on a device if the manufacturer cannot prove conformity to the regulation and that the device has been assessed in accordance with the appropriate procedures. The CE mark also means that the product can be freely marketed anywhere in the European Economic Area (EEA) without further control.

1.2. OVERVIEW OF THE CE MARKING PROCESS

There are a range of regulatory bodies involved in the CE marking process. These include competent authorities (CA), NBs, and the authorized representatives (AR). It is important for a manufacturer to understand the difference between each one. These are explained below:

1.2.1. Competent Authorities (CA)

A competent authority is defined as a regulatory body authorized by a European Member State government to monitor compliance of devices with the national legislation or regulations and carry out duties on behalf of the government in compliance with European Law. CA's existed in each European Member State and is nominated by each government to monitor and ensure compliance with the MDD (and upcoming MDR). The main role of the CA is to watch over the market in the member state for which it is designated. In particular, it must advise about the market launch of medical devices, advice about exporters and distributor's performance, advice about any clinical studies being done with medical devices in that particular territory, and advice about incidents that occur with medical devices. The CA designates a NB to ensure that conformity assessment procedures are completed as per relevant criteria.

A list of national CA within the EU is provided below:

Austria
Federal Ministry of Health-Department, Pharmaceuticals and Medical Devices, Radetzkystrasse – 2,1030, Wien/Vienna, Austria, E-mail: meddev@bmgf.gv.at
Federal Office for Safety in Health Care, Traisengasse – 5,1200, Vienna, Austria, Tel.: +43 50 555 36402, E-mail: medizinproduckte@basg.gv.at, Website: www.basg.gv.at

Belgium
Federal Agency for Nuclear Control, Health Protection Unit, Rue Ravenstein – 36, 1000, Brussels, Tel.: +32 2 289 21 11, Fax: +32 2 289 21 12, E-mail: pointcontact@fanc.fgov.be, Website: www.afcn.fgov.be
Federal Agency for Medicines and Health Products, Health Products Division, Place Victor Horta 40/40, 1060, Brussels, Tel.: +32 2 524 80 00, Fax: +32 2 524 81 20, E-mail: meddev@afmps.be, Website: www.fagg-afmps.be

Bulgaria
Bulgarian Drug Agency, Department Control of Medicinal Products and Medical Devices 8, Damyan Gruev Str., 1303 Sofia, Bulgaria, Tel.: +359 2 890 34 83, Fax:+359 2 890 34 34, E-mail: bda@bda.bg, Website: www.bda.bg

Croatia
Ministry of Health, Health Protection Directorate, Ksaver 200a, 10000 Zagreb, Croatia, Tel.: +385 1 4607555, Fax: +385 1 4677076, Website: www.zdravlje.hr
Agency for Medicinal Products and Medical Devices, Ksaverska Cesta 4, 10 000 Zagreb, Tel.: +385 1 4884 300, Fax: +381 1 4884 110, E-mail: halmed@halmed.hr, Website: www.halmed.hr

Cyprus
Ministry of Health, Cyprus Medical Devices Competent Authority 1 Prodromou and 17 Chilonos Street, 1448 Nicosia, Cyprus, Tel.: +357 22605572/735, Fax: +357 22468427, E-mail: cymda@mphs.moh.gov.cy, Website: www.moh.gov.cy/mphs

Czech Republic
State Institute for Drug Control, Šrobárova 48, 100 41 Praha 10, Tel.: +420 272 185 111, E-mail: posta@sukl.cz

Denmark
Danish Health and Medicines Authority, Axel Heides Gade 1 DK – 2300 Copenhagen S, Tel.: +45 72 22 74 00, Fax: +45 44 88 95 99, E-mails: med-udstyr@dkma.dk; dkma@dkma.dk, Website: www.medicinskudstyr.dk

Estonia
Health Board, Medical Devices Department, 1a Põllust., EE-Tartu – 50303, E-mail: MSO@terviseamet.ee, Website: www.terviseamet.ee

Finland
National Supervisory Authority for Welfare and Health (Valvira), Mannerheimintie 103 b, P.O. Box 210, FI – 00281 Helsinki, Finland, Tel.: +358 295 209 111, Website: www.valvira.fi

France
Ministry of the Economy and Finances, Direction Générale de la Concurrence, de la Consommation et de la Répression des Frauds-Unit 5B, Télédoc 241-59, Boulevard Vincent-Auriol F – 75703 Paris Cedex 13, Tel.: +33 1 44 97 28 01, E-mail: Bureau-5B@dgccrf.finances.gouv.fr, Website: www.economie.gouv.fr/dgccrf
Agence Nationale de Sécurité du Médicament et des Produits de Santé (ANSM), 143/147, Boulevard Anatole France 93285 Saint-Denis Cedex, Tel.: +33 1 55 87 36 86/+33 1 55 87 37 28, E-mails: dmdpt@ansm.sante.fr; dmtcos@ansm.sante.fr, Website: www.ansm.sante.fr
Direction Générale des Douanes et Droits Indirects, Tel.: 0811 20 44 44/+33 1 72 40 78 50, E-mail: ids@douane.finances.gouv.fr, Website: www.douane.gouv.fr

Germany
Federal Institute for Drugs and Medical Devices, Kurt Georg Kiesinger Allee 3, D – 53175 Bonn, Fax: +49 228 207 5300, E-mail: medizinprodukte@bfarm.de
Zentralstelle der Länder für Gesundheitsschutz bei Arzneimitteln und Medizinprodukten (ZLG), E-mail: zlg@zlg.nrw.de
Federal Ministry of Health, Rochusstrasse 1, D – 53123 Bonn, Fax:+ 49 228 941-4946, E-mail: medizinprodukte@bmg.bund.de

Greece
National Organization for Medicines, 284 Mesogion Ave, GR – 15562 Holargos, Athens E-mail: relation@eof.gr, Website: www.eof.gr

Hungary
Health Registration and Training Center (HRTC), H-1085 Budapest, Horánszky street 15, Tel.: +36 1 302 5060, Fax: +36 1 269 1255, E-mail: amd@enkk.hu, Website: www.enkk.hu

Iceland
Medicines Agency, Vínlandsleið 14, 113 Reykjavík, Tel.: 520 2100, Fax: 561 2170, E-mail: ima@ima.is, Website: www.imas.is

Ireland
Health Products Regulatory Authority (HPRA), Kevin O'Malley House, Earlsfort Centre, Earlsfort Terrace, Dublin 2, Tel.: +353 1 676 4971, E-mail: info@hpra.ie, Website: www.hpra.ie

Italy
Ministry of Health, Department of Planning and Organization of the National Health Service-Directorate, General of Medical Devices, Pharmaceutical Services, and Safety in Healthcare, via Giorgio Ribotta 5, IT – 00144 Roma, E-mail: segr.dgfdm@sanita.it, Website: www.sanita.it

Latvia
Health Inspectorate, Klijanu Street 7, Riga, LV—1012, Latvia, Tel.: 371-67819671, Mob: 371-26315189, Fax: 371-67819672, E-mail: vi@vi.gov.lv, Website: www.vi.gov.lv Liechtenstein, Amt Für Gesundheit, Äulestrasse 51, Postfach 684, 9490 Vaduz, Liechtenstein, E-mail: info.ag@llv.li, Website: www.ag.llv.li

Lithuania
State Health Care Accreditation Agency, Jeruzales St. 21 LT – 08420 Vilnius Lithuania, Tel.: (85) 261 5177, Fax: (85) 212 7310, E-mail: vaspvt@vaspvt.gov.lt, Website: www.vaspvt.gov.lt

Luxembourg
Ministère de la Santé, Villa Louvigny-Allée Marconi L – 2120 Luxembourg, Tel.: +352 24785500, E-mail: info@ms.public.lu

Malta
Malta Competition and Consumer Affairs Authority, Technical Regulations Division-Market Surveillance Directorate, Mizzi House, National Road, Blata l-Bajda, Hamrun, Malta HMR 9010, Tel.: +356 23952000, Fax: +356 21242406, E-mail: info@mccaa.org.mt, Website: www.mccaa.org.mt

Netherlands
The Health Care Inspectorate, Stadsplateau 1 3521 AZ Utrecht, Postbus 2680 3500 GR Utrecht, Tel.: +31 88 120 50 000, E-mail: meldpunt@igz.nl, Website: www.igz.nl

Norway
The Norwegian Medicines Agency, PO Box 240 Skøyen, 0213 Oslo, Norway, Tel.: +47 22 89 77 00, E-mail: meddev-no@noma.no, Website: www.legemiddelverket.no

Poland
Office for Registration of Medicinal Products, Medical Devices and Biocidal Products, Department of Medical, Devices Surveillance, Vigilance, and Clinical Trials, Al. Jerozolimskie 181C, 02-222 Warsaw, Poland, Tel.: +48 22 492 15 54, Fax: +48 22 492 11 09, E-mail: incydenty@urpl.gov.pl
Website: http://www.urpl.gov.pl/

Portugal
INFARMED-National Authority of Medicines and Health Products, IP, Parque de Saúde de Lisboa, Avenida do Brasil 53, 1749-004 Lisboa, Tel.: +351 217 987 100, Fax: +351 217 987 316, E-mail: infarmed@infarmed.pt, Website: www.infarmed.pt

Romania
National Agency for Medicines and Medical Devices, 58, Sos. Nicolae Titulescu, Sector 1, Bucharest, Tels.: +4 021 2228652, +4 031 432 9120, Fax: +4 021 2228683, +4 031 432 9121, E-mail: mdevice@anm.ro, Website: www.anm.ro

Slovak Republic
National Institute for Drug Control, Kvetná 11 825 08 Bratislava 26 Slovak Republic, Tel.: +421-2-5070 11 11, Fax: +421-2-55 56 41 27, E-mail: sukl@sukl.sk, Website: www.sukl.sk

Slovenia
Agency for Medicinal Products and Medicinal Services, Ptujska Ulica 21, 1000 Ljubljana, Slovenia, Tel.: +386 (0)8 2000 500, Fax: +386 (0)8 2000 630, E-mails: info@jazmp.si; med-dev@jazmp.si, Website: www.jazmp.si

Spain
Agencia Española de Medicamentos y Productos Sanitarios, C/ Campezo 1, Edificio 8, ES – 28022 Madrid, E-mail: sgps@aemps.es, Website: http://www.aemps.gob.es/

Sweden
The Medical Products Agency, P.O. Box 26, SE – 751 03 Uppsala, Tel.: +46 18 17 46 00, Fax: +46 18 50 31 15, E-mail: meddev.central@mpa.se, Website: www.lakemedelsverket.se
The Health and Social Care Inspectorate, P.O Box 45184, SE – 104 30 Stockholm, Tel.: +46 10-788 50 00, E-mail: registrator@ivo.se, Website: www.ivo.se

Switzerland
Swissmedic, Swiss Agency for Therapeutic Products, Hallerstrasse 7, CH – 3012 Berne, E-mail: medical.devices@swissmedic.ch, Website: https://www.swissmedic.ch

Turkey
Ministry of Health-Turkish Medicines and Medical Devices Agency, Vice Presidency of Inspectorate-Medical Devices Inspection Department, Söğütözü Mahallesi 2176. Sokak No:5 PK 06520 Çankaya/ANKARA, Tel.: +90 312 218 35 34, Fax: +90 312 218 37 10, E-mail: tibbicihazpgd@titck.gov.tr, Website: http://www.titck.gov.tr

United Kingdom
Medicines and Healthcare Products Regulatory Agency, 151 Buckingham Palace Road, London SW1W 9SZ, Tel.: +44 203 080 6000, Fax: +44 203 118 9803, E-mail: info@mhra.gsi.gov.uk

1.2.2. Notified Bodies (NB)

A Notified Body is defined as an independent certification organization that is "notified" by the competent authority of a European Member State. It plays an imperative role when it comes to supporting a manufacturer's ability to place safe and compliant medical devices on the European market. The notified body is important for determining if a product or system meets the applicable requirements of the specific legislation/for CE marking and assists with carrying out conformity assessment procedures to demonstrate that the product complies with the requirements of the legislation. There are specific NBs dedicated to performing conformity assessments under Directive 93/42/EEC however, it should also be noted that even fewer NBs are designated to perform conformity assessment under the Regulation (EU) 2017/745. Examples of these NBs are: BSI Assurance UK Ltd, BSI Group the Netherlands B.V., DEKRA Certification B.V, TUV SUD Rheinland and Intertek Medical Notified Body AB to name a few.

1.2.3. Authorized Representatives (AR)

The European AR (also known as the EU AR, EC REP, CE REP, EU REP, or EAR) serves as a liaison between a manufacturer and the national competent authority and NBs. The AR should be identified on the product labeling, outer packaging and the Instructions for Use (IFU) and will assist with device registrations. The AR will hold a copy of the manufacturer's technical file and/or CE Declaration of Conformity and will make it available to the CA upon request. They will also assist with Incident and Field Safety Corrective Action reporting. In many cases, the AR will also provide manufacturers with regulatory news and relevant updates.

The CA, NBs, and AR are all involved in the CE marking process which is interlinked between all three. The CA designates a NB to ensure that conformity assessment procedures are completed as per the relevant criteria. The authorized representative is designated by manufacturers and is legally responsible for compliance with the regulations. The AR acts as the first point of contact for the EU authorities. It is up to the manufacturer to ensure that their product complies with the essential requirements (ER) of the relevant EU legislation.

A general overview of the CE marking process is given in Figure 1.1.

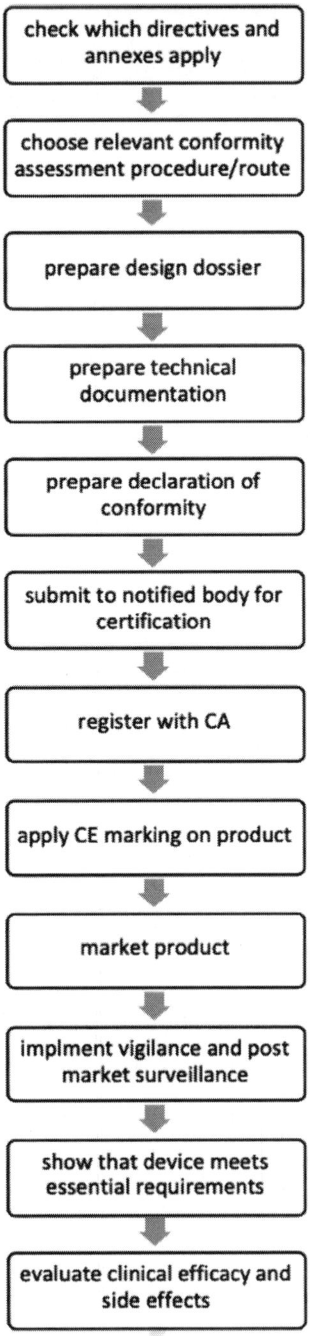

Figure 1.1. General overview of the CE marking process.

CHAPTER

2

The History of Medical Device Regulations

CONTENTS

2.1. HISTORY

Prior to the 1930's, the medical device industry was practically non-existent. Much of the medical devices being manufactured were sold directly to the public without any form of regulations to guide performance, safety, or functionality. After World War II, the medical device industry and related technologies grew exponentially leading the first developments of any kind of regulating between industry, government, and the public.

By the late 1960s, thousands of medical device manufacturers began to ship their devices. In fact, it was estimated that the total shipped products amounted to a value exceeding $1 billion. In the early 1970s, a released government report documented more than 10 000 injuries as a result of medical devices. Medical device regulations (MDRs) tend to vary a great deal from country to country. Over the last 20 or so years, the number, range, and complexity of medical devices has increased and as a result, concurrently, so has the complexity of the regulations involved.

The last twenty years or so has seen an explosion in the variety, use, and complexity of medical devices. These devices vary greatly in complexity, ranging from implantable devices such as pacemakers to less risky devices such as bedpans and gloves. In 2016 alone, the total global medical device industry was estimated at $130 billion with almost half of the global production and consumption occurring in the United States alone.

In 2001, the WHO published an "A Model Regulatory Program for Medical Devices," which served as an international guide to assist member states in establishing regulatory programs for medical devices.

The aim of this model was to provide information to nations that lacked a medical device regulatory system such that this guide would then enable the production of devices that met internationally compatible regulations. This was then followed by the release of 'Medical Devices regulations. Global overview and guiding principles' in 2003 by WHO. The main purpose of this guide was to emphasize the complexity of the medical device industry and identify issues that related to regulations.

This document essentially provides guidance to member states who want to create new, or modify existing medical device regulatory systems. By 2010, the WHO had then released an estimated report detailing numerically an overview of countries and related medical device regulatory frameworks (Table 2.1).

Table 2.1. Overview of Country Percentages and Related Medical Device Regulatory Frameworks

Countries (%)	Medical Device Regulations
30	Countries that have a developed framework for regulation of medical devices
30	Countries that partially have regulations of medical devices
40	Countries that are either developing a framework or do not yet have any regulation

2.2. INCIDENTS THAT LED INFLUENCED CHANGES TO THE REGULATORY FIELD FOR MEDICAL DEVICES

All around the world, medical devices have been involved in various negative incidents on patients which have either injured or, in worse cases, led to patient death. On the other hand, health authorities and various regulatory bodies across the globe have subsequently failed to protect such patients from these poorly functioning (and tested) devices which have had various negative effects on patients. These include the puncturing of organs, erroneous shocks, rotting of bones and organs, blood poisoning and other effects. A year-long investigation by the International Consortium of Investigative Journalists (ICIJ) in 2018 revealed that many governments actually hold many of these implants (regardless of the complexity or risk classification) to much lower safety testing standards as when compared to pharmaceuticals. Flaws within the regulatory field itself have resulted in medical device manufacturers leaving such problematic devices on the market in some countries while pulling them off the market in other countries. In fact, the Implant Files Investigation conducted by ICIJ which formed one of the biggest and first-ever global examination of the medical device industry revealed just how health and regulatory authorities all around the world have failed to protect millions of patients due to them being uninformed of the crucial risks around such devices. Many of these reported incidents and various past ones have highlighted the need for improvement across standards, processes, procedures, and overall legislation where medical devices are concerned. Many of these incidents have directly highlighted the need for strengthening existing legislation, specifically within the European Union which has, in turn led to drastic improvements and changes over the past decade ultimately giving regulators more control and oversight.

Previous European legislation adopted in the 1990s aimed at ensuring the smooth functioning of the internal market together with an attempt to maintain high health and safety levels. There have however been several key moments in industry that have highlighted the shortcomings of such legislative frameworks.

In April 2017, the new EU Medical Devices regulations and IVD regulations were adopted by the European Parliament. These new regulations entered into force on 25 May 2017 and stated that the fundamental revision of the legislation was to provide a 'sustainable regulatory framework for medical devices which ensure a high level of safety and health while supporting innovation.' The main purpose of the changed regulations was to ensure smooth functioning of the internal market with regards to medical devices and enforce a high level of health protection for patients and users.

2.2.1. The PIP Breast Implant Scandal

The PIP breast implant scandal was one of the biggest medical device industry scandals that ultimately influenced and resulted in major regulatory updates, not only within the EU but on a global level. Medical devices in the European Union are regulated by the Medical Devices Directive which had also been transposed in France in several provisions of the French Public Health Code and in England in the Medical Devices Regulations (2002).

In 1965, plastic surgeon Henri Arion introduced breast implants to France. Together with Jean-Claude Mas, they launched a Poly Implant Prothese (PIP) company in 1991. This company essentially produced a figure or around 2 million sets of silicone breast implants over a 20-year period, however despite their success; they also created one of the biggest global health scares. This changed the medical device regulatory field.

The health scare came about when PIP decided to change the material being used to manufacture the silicone breast implants to a cheaper, industrial-grade silicone that was not approved for medical use. Industrial grade silicone is thought to have more contaminants that medical grade silicone however, it is thought the company chose to use the alternative to save costs and boost overall profits. Investigations later revealed that this industrial grade silicone had a rupture rate that was double that of the general industry average. In addition, later studies revealed that if compromised, the silicone gel was thought to cause both, short and long-term effects such as inflammation and possible scarring in affected patients.

In December 2011, French authorities issued a worldwide alert, advising all women who had been fitted with potentially defective implants to have them removed. By this time, PIP had become the world's third largest supplier of implants with over 300 000 women in 65 different countries around the world having received the implants, be they for cosmetic reasons or medical reasons such as reconstructive surgery after breast cancer treatment. Much of the implants had been used widely across Europe. It is estimated that around 40 000 women in Britain and South America alone may have received such implants.

Much of the advice given to women from regulatory bodies and surgeons alike varied from country to country. In France, women who had received such implants were urged to have them removed as a precaution. In Venezuela, regulatory authorities further recommended removal while the British government on the other hand said there was no evidence to recommend routine removal of PIP implants. They did however advise women who were concerned to have them removed. In Germany, medical groups advised women to seek removal of their implants with a further statement indicating that there was no immediate danger.

The French government subsequently offered to pay for the removal and replacement of all PIP implants fitted in France while Venezuela said it would cover the cost of implant removals but not the replacements. Within the United Kingdom, those women who had their implants paid for by the National Health Service (NHS) was able to have them removed free of charge after a consultation with their doctor while private companies who had fitted PIP implants for cosmetic purposes were advised to follow suite.

2.2.2. Overview of Events in the PIP Scandal (Table 2.2)

Table 2.2. Overview of the Events in the PIP Scandal

1991	• PIP was launched by Jean-Claude Mas
1992	• The United States Food and Drug Administration (FDA) calls for voluntary moratorium on use of silicone gel Implants. Implant sales are halted in the United States.
1994	• PIP begins selling hydrogel implants.
1996	• PIP begins selling saline-filled implants in the United States.
1997	• PIP authorized to produce medical-grade silicone implants.
2000	• The FDA refuses to approve PIP's saline-filled implants and warns about deviations from good manufacturing practices (GMP) found at the PIP plant. The company withdraws its hydrogel implants from the market as it cannot show they are safe.

2001	• PIP starts using an unapproved industrial-grade silicone in their implants.
2003	• Regulatory bodies begin to trace the first signs of legal problems and financial losses.
2009	• Surgeons started reporting abnormally high rupture rates in France sparking concern. This led to a host of legal complaints and PIP undergoing bankruptcy. The UK regulatory authority subsequently warned of several medical claims being made.
2010	• PIP was placed into liquidation after the French medical safety agency recalled its implants.
2011	• The French government recommended that approximately 30 000 women with PIP implants seek removal of the implants as a health precaution.
2012	• Jean-Claude Mas, the founder of PIP is arrested. France launches an investigation into the scandal.
2013	• Jean-Claude Mas is sent to prison for four years and fined 75 000 Euros. He is subsequently banned for life from working in medical services or running a company. He also faced two additional legal cases: one for involuntary manslaughter and another linked to the financial implications of the scandal.
2016	• Mas's sentence upheld by appeal court.
2017	• New European Medical Devices Regulation (MDR) published and in force, with transition period for manufacturers to comply over 3 years.

2.2.3. What Happened to PIP?

PIP implants were banned and eventually the company went bankrupt in 2010. The founder of PIP, Jean Claude Mas went under investigation by the French police and was subsequently charged with involuntary injury over the implants. PIP's lawyer eventually did respond to the scandal stating that industrial silicone was used in their implants and that while no tests conducted showed any danger, the main issue was public fear which subsequently created intense media pressure and pushed the blame ultimately onto PIP. Regardless, the concerns raised over the PIP implants have led to one of the biggest calls for changes to regulations within the plastic surgery industry. Ultimately, the consumer does not know what should be regarded as being safe or not hence then need for more stringent regulations.

The events of the PIP scandal resulted in the European Commission seriously reviewing and changing their MDRs. In fact, the regulation of the medical device industry was already in the process of being reformed when the PIP scandal occurred. The scandal however increased the focus and need for such change. In 2008, the European Commission published a public consultation document that asked stakeholders for their view around the revision of the legal framework of medical devices. This was followed by a public consultation in 2010 which focused on the technical aspects related to the revision of Directive 98/79/EC on *in vitro* diagnostic medical devices.

2.2.4. Changes to the Medical Device Regulations (MDRs) as a Result of PIP

Ultimately the main reasons for such an industry overall were that, at the time, existing EU regulations which dated back to the 1990's had not kept pace with the technological progression over the years. This had resulted in many regulatory gaps and uncertainties around products being manufactured from non-viable human tissues or cells as well as implantable and/or other invasive products used for cosmetic purposes. Furthermore, there were substantial divergences in the overall interpretation and application of the regulations/rules that emerged between EU countries. Much of this varied interpretation arose from each country having to interpret and respond to such technological and scientific progression individually. In addition, there was limited ability to trace medical devices back to their suppliers.

Essentially the new regulations now referred to as the European Medical Devices Regulation (MDR) which also includes regulations (EU) 2017/745 and 2017/746 have become far more stringent. Some of the major changes include:

- Expansion of the overall scope within the MDR to cosmetic and/or aesthetic devices.
- Clarification regarding areas such as medical software and generic tests.
- Reflection of technological advances through the adaptation of safety and performance requirements that are applicable to new health technologies.
- Stricter requirements for medical device compliance from manufacturers to ensure overall patient safety.
- Stronger supervision of independent assessment bodies by national authorities.
- More power and obligations for those above-mentioned assessment bodies to ensure thorough testing and regular checks on manufacturers such as unannounced factory inspections, sample testing.

- Improvement of co-ordination between national surveillance bodies.
- The creation of a national registry to increase traceability. This allows manufacturers to register themselves and the devices they place on the EU market within a central database through a unique device identification system. This way it will make it easier to trace and identify devices and enhance overall post-market safety.
- A ban on aggressive marketing of medical devices.
- Overall requirements for improved training and qualification standards.
- Increased and improved transparency along the entire supply chain of medical devices.
- The establishment of a portal within the European Union, which requires manufacturers to report serious events and corrective actions to reduce the overall risk of such events happening again.
- An increase in the rights and responsibilities of notified bodies (NBs) as well as the stringency with which they monitor all medical devices.
- A further requirement that within a manufacturers organization to have a qualified person responsible for regulatory compliance.
- Patients that have been implanted with a device should also be given essential information which will allow the device to be identified. This information should also contain necessary warnings or precautions that must be taken.

2.2.5. The DePuy Hip Replacement Recall

MoM hip designs were first introduced in the 1950s and 1960s until such time that improvements in manufacturing and measurement technologies sparked designers to revisit the overall design of MoM hips in the 1990s. In 2003, the medical implant manufacturer DuPuy Orthopedics which is a subsidiary of Johnson and Johnson introduced their own version of a metal-on-metal (MoMP) hip resurfacing Articular Surface Replacement (ASR) ASRTM and ASRTM XL, a total hip replacement version. Both versions were available in Europe and were available under the clause of 'substantial equivalence' which, allows for fast-track market accreditation through claiming the implant in question is similar to ones already on the market.

As per regulatory bodies, the ASRTM hip resurfacing was granted market approval by the FDA in the USA while the ASRTM XL was available globally.

In 2008, problems with ASRTM were first raised at various orthopedic conferences as well as in journal publications.

This was followed by an annual report published by the National Joint registry of England and Wales in 2009 which showed high revision rates of approximately 7.5% at three years for the ASRTM hip resurfacing.

At the time, however, official notices pertaining to such problems were limited to guidance on positioning during surgery until around August 2010 when DuPuy Orthopedics issued a global recall of all ASRTM and ASRTM XL hip systems.

In addition, the Medicines and Healthcare products Regulatory Agency (MHRA) further issued a general Medical Device Alert (MDA) on 22 April 2010 for all metal-on-metal (MoM) hip replacements (MDA/2010/033).

This MDA did not specifically address the ASRTM implants but rather showed that much data had been available for a while indicating that general concerns around MoM hip implants did exist. It should also be noted that prior to this, high revision rates with ASRTM were also highlighted in Australia somewhere in 2007. As a consequence of this, the ASRTM was withdrawn at the end of 2009.

2.2.6. Overview of Events in the DuPuy Scandal

The impact of such high revision rates on patients is significant and should be noted. Failed MoM hip implants can cause both local and systemic health problems as well as symptoms such as hearing, loss, dizziness, decline in cognitive functions, cardiomyopathy, and organ failure while ultimately could result in increased pain and a decrease in overall mobility.

Various reasons were attributed to failure of the hip replacement system. These include loosening and misalignment of components, infection, and fracture of the bone, dislocation, metal sensitivity, and overall pain.

In addition, further complications could include an increase in the level of metal ions in the blood, staining of the bone, necrosis, swelling, and unknown damage to the nerves, tissues, and/or muscles (Table 2.3).

Table 2.3. Overview of Events in the DuPuy Scandal

2003	• ASR™ resurfacing introduced.
2007	• High revision rates reported by the Australian Orthopedic Association, National Joint Replacement Registry.
2008	• Subsequent reporting of high revision rates by Australian Orthopedic Association, National Joint Replacement Registry.
2009	• First report of high revision rates in the National Joint Registry of England and Wales. • ASR™ withdrawn in Australia and New Zealand.
2010	• MHRA issues MDA on all MoM implants noting a small number of patients have adverse reactions. • DePuy release guidance on positioning. • MHRA issue MDA on positioning. • DePuy withdraw the ASR™ globally.

2.2.7. Reactions to Published Data on DuPuy Hip Replacement Scandal

In 2007, the Australian Orthopaedic Association national joint replacement registry noted that ASR seemed to have twice the risk of revision thus indicating the need for further surgery as when compared with other resurfacing procedures. This was followed by the registry publishing data the next year showing that some women undergoing metal-on-metal hip resurfacing were facing more follow-up operations.

At the beginning of 2010, DuPuy Orthopedics issued a statement saying they were phasing out the ASR Hip Implant due to declining sales however; they failed to mention the data from the Australian implant registry. Later, sometime in March 2010, DuPuy issued their first warning to doctors and patients indicating the high early failure rate but still had not issued a product recall. Eventually DuPuy Orthopedics withdrew their implants from the market after the National Joint registry (NJR) in England and Wales shared data that showed patients with their MoM hips having a higher expected revision rate. In fact, data showed that within a five-year period of receiving DuPuy's resurfacing device, 12% of patients had undergone a revision surgery and within five years of having an ASR total hip replacement, 13% of patient had undergone revision surgery.

It was estimated that about 93 000 people globally had received a DuPuy implant. The first lawsuit against DuPuy Orthopedics was filed in the United States on June 15, 2010 followed another on September

27, 2011. Both lawsuits claimed that DePuy ASR hip replacements were improperly designed and that DePuy knew there were problems with their implants early on and despite this, did not alert patients or surgeons about the possible problems. In 2014, Johnson, and Johnson announced that it would not withdraw a $2.5 billion global settlement. Settlements began at a base of $250,000. Johnson and Johnson continued to deny culpability. It should be further noted that of the total number of people known to receive the implants, approximately 4,700 were thought to be from India. Johnson and Johnson then undertook to working with the Indian government in 2018 in an attempt to support all Indian patients affected by the implants.

2.2.8. Inadequate Vaginal Mesh

Back in the late 1980s, the medical industry was searching for new ways to treat women that were experiencing urinary incontinence as well as vaginal prolapsed. Both are common conditions that women experience after childbirth and while most doctors, at the time, suggested physiotherapy, weight-loss, and other non-surgical interventions as a solution; the use of mesh implants seemed like a simple, quick, and convenient alternative. The mesh was initially designed to allow a patient's body tissues to grow through it, making it harder to remove. Mesh implants took about an hour to implant and allowed women to leave hospital quickly and soon became the standard treatment for millions of women all around the world. While many cases have had effective results, there are also some women who have experienced severe complications such as eroding mesh that has pierced the vaginal wall or bladder, nerve damage and infections. In addition, many of these women have experienced chronic pain and in some cases were barely able to walk.

One of the largest manufacturers of gynecological mesh is medical device manufacturer, Boston Scientific. While the company has been successful in the past, they have also attracted close to 48 000 lawsuits against them, all of which claim that the gynecological mesh used can cause extreme pain and injury. Despite such claims, Boston Scientific has continued to fight such allegations with a comeback stating that over 1 million women have been successfully treated and that they have extensively tested the plastic/resin used to confirm the mesh composition, safety, and performance. Unfortunately, many doctors have also found otherwise, with some claiming the mesh causes a chronic inflammatory reaction and stating that once removed; the mesh seems to have a substantially altered architecture. In addition, they seem to have shrunk by 50% in width and are excised in scar tissue.

Investigations revealed that the mesh is made of a polypropylene plastic which is known to be a common packaging material. Boston Scientific however has proof that they have clearance from the FDA to use a particular brand of polypropylene called Marlex. This brand is said to be made in Texas by a subsidiary of Chevron Phillips. In 2004 however, Chevron Phillips became concerned with the use of Marlex in the medical industry and issued a warning that it must not be used for any form of permanent implantation in the body. This is due to the unstable oxidative state of polypropylene. Oxygen is known to break up the polypropylene and cause dissipation of antioxidant additives over time and as a result, causes the polypropylene to fall apart within the body. As time passed, Chevron Phillips cut off Boston Scientifics' supply of Marlex, somewhere in 2005; however, Boston Scientific remained adamant on continuing with their processes. As a result, they continued their search for a supplier and eventually came across a broker in China who claimed to have large amounts of Marlex that was supposedly imported from Chevron Phillips in Texas. Simultaneously that same month, the FDA issued a report claiming that over a five-year period, they had found about 4000 reports of injury, death, and malfunction as well as associated complications including pain, infection, urinary problems, bleeding, and organ perforation. This further gave industry a reason to believe that if they planned on switching plastics, the FDA would now require years of testing which ultimately might fail and hinder sales.

The Chinese company seemed like the best option at the time. Boston Scientific had established their own procedures for what required documents would be needed as well as requests for various import records to prove the Marlex was really imported from Texas by the Chinese company. Unfortunately, the Chinese company did not have any documents that could ensure the authenticity of the Marlex could be verified. Further investigation revealed that lot numbers on the bags did not match up as well as counterfeit printing. Boston Scientific then further ordered testing on the material and analyzed 11 different parameters. Results from nine of these tests differed. Despite this, Boston Scientific still bought a large amount of Chinese plastic and continued to use it in their manufacturing processes.

CHAPTER

3

The Medical Device Directive (MDD)

CONTENTS

3.1. WHAT IS THE MEDICAL DEVICE DIRECTIVE (MDD)?

Any medical device placed on the market must comply with the relevant legislation of the country in which that device is entering. For Europe, this was the medical device directive (MDD); in transition to the medical device regulation (MDR). Thismeansn, for a manufacturer to legally place a medical device on the European market, they must show that their device meets the essential requirements (ER) of the relevant regulation.

The legislation consists of three directives:

- Directive 90/385/EEC for active implantable medical devices (AIMD);
- Directive 93/42/EEC for medical devices (MDD); and
- Directive 98/79/EC for *in vitro* diagnostic medical devices (IVDD).

As such, all devices being sold within the European Union were required to conform to the directive by June 1, 1998. It's important to understand that the main purpose of the MDD was to harmonize those standards that benefit manufacturers, users, and patients while also defining the specific requirements for clinical testing, design, manufacture, testing/inspection, marketing, installation, and service of medical devices being sold within the European Union.

3.2. UNDERSTANDING THE MDD: LET'S TAKE A CLOSER LOOK

The MDD consists of 23 articles, 12 annexes, and 18 classification rules. Within the MDD, three main areas of focus for medical devices were:

1. The essential requirements;
2. Classification rules; and
3. Conformity routes for assessments.

3.3. ESSENTIAL REQUIREMENTS (ER)

This section of the MDD focuses on the minimum ER for the design and manufacture of medical devices. It is aimed at ensuring the protection of the health and safety of patients, users, and third parties. For this reason, these requirements stipulate that the principles of safety should be integral to the design of the product and that the product should be suitable for its intended purpose. The ER are defined in Annex I of the MDD.

Annex 1 of the Directive lays down a series of ER for the design, safety, and functionality of medical devices. The ER of the MDD apply to all medical devices and cover the following:

- A general requirement for safe design;
- The minimization of risks from contamination;
- Compatibility with materials with which they are likely to come into contact;
- Minimization of hazards of infection and microbial contamination;
- Provision of sufficient accuracy (for devices with a measuring function);
- Protection against radiation;
- Adequate product marking;
- Adequate user instructions.

The specific details and requirements of these ER are laid out in the directive and are further defined below:

Excerpt from the MDD: Article 3: Essential Requirements.

Article 3 of the MDD 93/42/EEC: Essential Requirements
The devices must meet the ER set out in Annex I which apply to them, taking account of the intended purpose of the devices concerned. *Where a relevant hazard exists, devices which are also machinery within the meaning of Article 2(a) of Directive 2006/42/EC of the European Parliament and of the Council of 17 May 2006 on machinery (5) shall also meet the essential health and safety requirements set out in Annex I to that Directive to the extent to which those essential health and safety requirements are more specific than the ER set out in Annex I to this Directive.*

Now let's look at the requirements of Annex I (Essential Requirements) in greater detail.

Annex I of the MDD 93/42/EEC: Essential requirements

I. General Requirements

1. The devices must be designed and manufactured in such a way that, when used under the conditions and for the purposes intended, they will not compromise the clinical condition or the safety of patients, or the safety and health of users or, where applicable, other persons, provided that any risks which may be associated with their intended use constitute acceptable risks when weighed against the benefits to the patient and are compatible with a high level of protection of health and safety.

This shall include:

- *Reducing, as far as possible, the risk of use error due to the ergonomic features of the device and the environment in which the device is intended to be used (design for patient safety).*
- *Consideration of the technical knowledge, experience, education, and training and where applicable the medical and physical conditions of intended users (design for lay, professional, disabled or other users).*

2. The solutions adopted by the manufacturer for the design and construction of the devices must conform to safety principles, taking account of the generally acknowledged state of the art.

In selecting the most appropriate solutions, the manufacturer must apply the following principles in the following order:

- *Eliminate or reduce risks as far as possible (inherently safe design and construction).*
- *Where appropriate take adequate protection measures including alarms if necessary, in relation to risks that cannot be eliminated.*
- *Inform users of the residual risks due to any shortcomings of the protection measures adopted.*

3. The devices must achieve the performances intended by the manufacturer and be designed, manufactured, and packaged in such a way that they are suitable for one or more of the functions referred to in Article 1 (2) (a), as specified by the manufacturer.

4. The characteristics and performances referred to in Sections 1, 2 and 3 must not be adversely affected to such a degree that the clinical conditions and safety of the patients and, where applicable, of other persons are compromised during the lifetime of the device as indicated by the manufacturer, when the device is subjected to the stresses which can occur during normal conditions of use.

5. The devices must be designed, manufactured, and packed in such a way that their characteristics and performances during their intended use will not be adversely affected during transport and storage taking account of the instructions and information provided by the manufacturer.

6. Any undesirable side-effect must constitute an acceptable risk when weighed against the performances intended.

6a. Demonstration of conformity with the ER must include a clinical evaluation in accordance with Annex X.

II. Requirements Regarding Design and Construction

7. Chemical, Physical, and Biological Properties.

7.1. The devices must be designed and manufactured in such a way as to guarantee the characteristics and performances referred to in Section I on the 'General requirements. Particular attention must be paid to:

- *The choice of materials used, particularly as regards toxicity and, where appropriate, flammability.*
- *The compatibility between the materials used and biological tissues, cells, and body fluids, taking account of the intended purpose of the device.*
- *Where appropriate, the results of biophysical or modeling research whose validity has been demonstrated beforehand.*

7.2. The devices must be designed, manufactured, and packed in such a way as to minimize the risk posed by contaminants and residues to the persons involved in the transport, storage, and use of the devices and to the patients, taking account of the intended purpose of the product. Particular attention must be paid to the tissues exposed and to the duration and frequency of exposure.

7.3. The devices must be designed and manufactured in such a way that they can be used safely with the materials, substances, and gases with which they enter into contact during their normal use or during routine procedures; if the devices are intended to administer medicinal products they must be designed and manufactured in such a way as to be compatible with the medicinal products concerned according to the provisions and restrictions governing these products and that their performance is maintained in accordance with the intended use.

7.4. Where a device incorporates, as an integral part, a substance which, if used separately, may be considered to be a medicinal product as defined in Article 1 of Directive 2001/83/EC and which is liable to act upon the body with action ancillary to that of the device, the quality, safety, and usefulness of the substance must be verified by analogy with the methods specified in Annex I to Directive 2001/83/ EC.

For the substances referred to in the first paragraph, the notified body shall, having verified the usefulness of the substance as part of the medical device and taking account of the intended purpose of the device, seek a scientific opinion from one of the competent authorities (CA) designated by the Member States or the European Medicines Agency (EMEA) acting particularly through its committee in accordance with Regulation (EC) No 726/2004 (1) on the quality and safety of the substance including the clinical benefit/risk profile of the incorporation of the substance into the device. When issuing its opinion, the competent authority or the EMEA shall take into account the manufacturing process and the data related to the usefulness of incorporation of the substance into the device as determined by the notified body. Where a device incorporates, as an integral part, a human blood derivative, the notified body shall, having verified the usefulness of the substance as part of the medical device and taking into account the intended purpose of the device, seek a scientific opinion from the EMEA, acting particularly through its committee, on the quality and safety of the substance including the clinical benefit/risk profile of the incorporation of the human blood derivative into the device. When issuing its opinion, the EMEA shall take into account the manufacturing process and the data related to the usefulness of incorporation of the substance into the device as determined by the notified body.

Where changes are made to an ancillary substance incorporated in a device, in particular related to its manufacturing process, the notified body shall be informed of the changes and shall consult the relevant medicines competent authority (i.e., the one involved in the initial consultation), in order to confirm that the quality and safety of the ancillary substance are maintained. The competent authority shall take into account the data related to the usefulness of incorporation of the substance into the device as determined by the notified body, in order to ensure that the changes have no negative impact on the established benefit/risk profile of the addition of the substance in the medical device.

When the relevant medicines competent authority (i.e., the one involved in the initial consultation) has obtained information on the ancillary substance, which could have an impact on the established benefit/risk profile of the addition of the substance in the medical device, it shall provide the notified body with advice, whether this information has an impact on the established benefit/risk profile of the addition of the substance in the medical device or not. The notified body shall take the updated scientific opinion into account in reconsidering its assessment of the conformity assessment procedure.

7.5. The devices must be designed and manufactured in such a way as to reduce to a minimum the risks posed by substances leaking from the device. Special attention shall be given to substances which are carcinogenic, mutagenic or toxic to reproduction, in accordance with Annex I to Council Directive 67/548/EEC of 27 June 1967 on the approximation of laws, regulations, and administrative provisions relating to the classification, packaging, and labeling of dangerous substances (1).

If parts of a device (or a device itself) intended to administer and/or remove medicines, body liquids or other substances to or from the body, or devices intended for transport and storage of such body fluids or substances, contain phthalates which are classified as carcinogenic, mutagenic or toxic to reproduction, of category 1 or 2, in accordance with Annex I to Directive 67/548/EEC, these devices must be labeled on the device itself and/or on the packaging for each unit or, where appropriate, on the sales packaging as a device containing phthalates.

If the intended use of such devices includes treatment of children or treatment of pregnant or nursing women, the manufacturer must provide a specific justification for the use of these substances with regard to compliance with the ER, in particular of this paragraph, within the technical documentation and, within the instructions for use,t information on residual risks for these patient groups and, if applicable, on appropriate precautionary measures.

7.6. Devices must be designed and manufactured in such a way as to reduce, as much as possible, risks posed by the unintentional ingress of substances into the device taking into account the device and the nature of the environment in which it is intended to be used.

8. Infection and microbial contamination.

8.1. The devices and manufacturing processes must be designed in such a way as to eliminate or reduce as far as possible the risk of infection to the patient, user, and third parties. The design must allow easy handling and, where necessary, minimize contamination of the device by the patient or vice versa during use.

8.2. Tissues of animal origin must originate from animals that have been subjected to veterinary controls and surveillance adapted to the intended use of the tissues.

NBs shall retain information on the geographical origin of the animals.

Processing, preservation, testing, and handling of tissues, cells, and substances of animal origin must be carried out so as to provide optimal security. In particular, safety with regard to viruses and other transmissible agents must be addressed by implementation of validated methods of elimination or viral inactivation in the course of the manufacturing process.

8.3. Devices delivered in a sterile state must be designed, manufactured, and packed in a non-reusable pack and/or according to appropriate procedures to ensure that they are sterile when placed on the market and remain sterile, under the storage and transport conditions laid down, until the protective packaging is damaged or opened.

8.4. Devices delivered in a sterile state must have been manufactured and sterilized by an appropriate, validated method.

8.5. Devices intended to be sterilized must be manufactured in appropriately controlled (e. g. environmental) conditions.

8.6. Packaging systems for non-sterile devices must keep the product without deterioration at the level of cleanliness stipulated and, if the devices are to be sterilized prior to use, minimize the risk of microbial contamination; the packaging system must be suitable taking account of the method of sterilization indicated by the manufacturer.

8.7. The packaging and/or label of the device must distinguish between identical or similar products sold in both sterile and non-sterile condition.

9. Construction and environmental properties.

9.1. If the device is intended for use in combination with other devices or equipment, the whole combination, including the connection system must be safe and must not impair the specified performances of the devices. Any restrictions on use must be indicated on the label or in the instructions for use.

9.2. Devices must be designed and manufactured in such a way as to remove or minimize as far as is possible:

- *The risk of injury, in connection with their physical features, including the volume/pressure ratio, dimensional, and where appropriate ergonomic features.*
- *Risks connected with reasonably foreseeable environmental conditions, such as magnetic fields, external electrical influences, electrostatic discharge, pressure, temperature, or variations in pressure and acceleration.*
- *The risks of reciprocal interference with other devices normally used in the investigations or for the treatment given.*
- *Risks arising where maintenance or calibration are not possible (as with implants), from aging of materials used or loss of accuracy of any measuring or control mechanism.*

9.3. Devices must be designed and manufactured in such a way as to minimize the risks of fire or explosion during normal use and in single fault condition. Particular attention must be paid to devices whose intended use includes exposure to flammable substances or to substances which could cause combustion.

10. Devices with a measuring function:

10.1. Devices with a measuring function must be designed and manufactured in such a way as to provide sufficient accuracy and stability within appropriate limits of accuracy and taking account of the intended purpose of the device. The limits of accuracy must be indicated by the manufacturer.

10.2. The measurement, monitoring, and display scale must be designed in line with ergonomic principles, taking account of the intended purpose of the device.

10.3. The measurements made by devices with a measuring function must be expressed in legal units conforming to the provisions of Council Directive 80/181/EEC (1).

11. Protection against radiation.

11.1. General.

11.1.1. Devices shall be designed and manufactured in such a way that exposure of patients, users, and other persons to radiation shall be reduced as far as possible compatible with the intended purpose, whilst not restricting the application of appropriate specified levels for therapeutic and diagnostic purposes.

11.2. Intended radiation.

11.2.1. Where devices are designed to emit hazardous levels of radiation necessary for a specific medical purpose the benefit of which is considered to outweigh the risks inherent in the emission, it must be possible for the user to control the emissions. Such devices shall be designed and manufactured to ensure reproducibility and tolerance of relevant variable parameters.

11.2.2. Where devices are intended to emit potentially hazardous, visible, and/ or invisible radiation, they must be fitted, where practicable, with visual displays and/or audible warnings of such emissions.

11.3. Unintended radiation.

11.3.1. Devices shall be designed and manufactured in such a way that exposure of patients, users, and other persons to the emission of unintended, stray or scattered radiation is reduced as far as possible.

11.4. Instructions.

11.4.1. The operating instructions for devices emitting radiation must give detailed information as to the nature of the emitted radiation, means of protecting the patient and the user and on ways of avoiding misuse and of eliminating the risks inherent in installation.

11.5. Ionizing radiation.

11.5.1. Devices intended to emit ionizing radiation must be designed and manufactured in such a way as to ensure that, where practicable, the quantity, geometry, and quality of radiation emitted can be varied and controlled taking into account the intended use.

11.5.2. Devices emitting ionizing radiation intended for diagnostic radiology shall be designed and manufactured in such a way as to achieve appropriate image and/or output quality for the intended medical purpose whilst minimizing radiation exposure of the patient and user.
11.5.3. Devices emitting ionizing radiation, intended for therapeutic radiology shall be designed and manufactured in such a way as to enable reliable monitoring and control of the delivered dose, the beam type, and energy and where appropriate the quality of radiation.
12. Requirements for medical devices connected to or equipped with an energy source.
12.1. Devices incorporating electronic programmable systems must be designed to ensure the repeatability, reliability, and performance of these systems according to the intended use. In the event of a single fault condition (in the system), appropriate means should be adopted to eliminate or reduce as far as possible consequent risks.
12.1a For devices which incorporate software or which are medical software in themselves, the software must be validated according to the state of the art taking into account the principles of development lifecycle, risk management, validation, and verification.
12.2. Devices where the safety of the patients depends on an internal power supply must be equipped with a means of determining the state of the power supply.
12.3. Devices where the safety of the patients depends on an external power supply must include an alarm system to signal any power failure.
12.4. Devices intended to monitor one or more clinical parameters of a patient must be equipped with appropriate alarm systems to alert the user of situations which could lead to death or severe deterioration of the patient's state of health.
12.5. Devices must be designed and manufactured in such a way as to minimize the risks of creating electromagnetic fields which could impair the operation of other devices or equipment in the usual environment.
12.6. Protection against electrical risks.
Devices must be designed and manufactured in such a way as to avoid, as far as possible, the risk of accidental electric shocks during normal use and in single fault condition, provided the devices are installed correctly.
12.7. Protection against mechanical and thermal risks.
12.7.1. Devices must be designed and manufactured in such a way as to protect the patient and user against mechanical risks connected with, for example, resistance, stability, and moving parts.
12.7.2. Devices must be designed and manufactured in such a way as to reduce to the lowest possible level the risks arising from vibration generated by the devices, taking account of technical progress and of the means available for limiting vibrations, particularly at source, unless the vibrations are part of the specified performance.
12.7.3. Devices must be designed and manufactured in such a way as to reduce to the lowest possible level the risks arising from the noise emitted, taking account of technical progress and of the means available to reduce noise, particularly at source, unless the noise emitted is part of the specified performance.

12.7.4. Terminals and connectors to the electricity, gas or hydraulic and pneumatic energy supplies which the user has to handle must be designed and constructed in such a way as to minimize all possible risks.
12.7.5. Accessible parts of the devices (excluding the parts or areas intended to supply heat or reach given temperatures) and their surroundings must not attain potentially dangerous temperatures under normal use.
12.8. Protection against the risks posed to the patient by energy supplies or substances.
12.8.1. Devices for supplying the patient with energy or substances must be designed and constructed in such a way that the flow-rate can be set and maintained accurately enough to guarantee the safety of the patient and of the user.
12.8.2. Devices must be fitted with the means of preventing and/or indicating any inadequacies in the flow-rate which could pose a danger. Devices must incorporate suitable means to prevent, as far as possible, the accidental release of dangerous levels of energy from an energy and/or substance source.
12.9. The function of the controls and indicators must be clearly specified on the devices.
Where a device bears instruction required for its operation or indicates operating or adjustment parameters by means of a visual system, such information must be understandable to the user and, as appropriate, the patient.
13. Information supplied by the manufacturer.
13.1. Each device must be accompanied by the information needed to use it safely and properly, taking account of the training and knowledge of the potential users, and to identify the manufacturer.
This information comprises the details on the label and the data in the instructions for use.
As far as practicable and appropriate, the information needed to use the device safely must be set out on the device itself and/or on the packaging for each unit or, where appropriate, on the sales packaging. If individual packaging of each unit is not practicable, the information must be set out in the leaflet supplied with one or more devices. Instructions for use must be included in the packaging for every device. By way of exception, no such instructions for use are needed for devices in Class I or IIa if they can be used safely without any such instructions.
13.2. Where appropriate, this information should take the form of symbols. Any symbol or identification color used must conform to the harmonized standards. In areas for which no standards exist, the symbols and colors must be described in the documentation supplied with the device.
13.3. The label must bear the following particulars:
a. The name or trade name and address of the manufacturer. For devices imported into the Community, in view of their distribution in the Community, the label, or the outer packaging, or instructions for use, shall contain in addition the name and address of the authorized representative where the manufacturer does not have a registered place of business in the Community;
b. The details strictly necessary to identify the device and the contents of the packaging especially for the users;

c. Where appropriate, the word 'STERILE';
d. Where appropriate, the batch code, preceded by the word 'LOT,' or the serial number;
e. Where appropriate, an indication of the date by which the device should be used, in safety, expressed as the year and month;
f. Where appropriate, an indication that the device is for single use. A manufacturer's indication of single use must be consistent across the Community;
g. If the device is costume-made, the words 'custom-made device';
h. If the device is intended for clinical investigations, the words 'exclusively for clinical investigations';
i. Any special storage and/or handling conditions;
j. Any special operating instructions;
k. Any warnings and/or precautions to take;
l. Year of manufacture for active devices other than those covered by (e). This indication may be included in the batch or serial number;
m. Where applicable, method of sterilization;
n. In the case of a device within the meaning of Article 1(4a), an indication that the device contains a human blood derivative.
13.4. If the intended purpose of the device is not obvious to the user, the manufacturer must clearly state it on the label and in the instructions for use.
13.5. Wherever reasonable and practicable, the devices and detachable components must be identified, where appropriate in terms of batches, to allow all appropriate action to detect any potential risk posed by the devices and detachable components.
13.6. Where appropriate, the instructions for use must contain the following particulars:
a. The details referred to in Section 13.3, with the exception of (d) and (e);
b. The performances referred to in Section 3 and any undesirable side-effects;
c. If the device must be installed with or connected to other medical devices or equipment in order to operate as required for its intended purpose, sufficient details of its characteristics to identify the correct devices or equipment to use in order to obtain a safe combination;
d. All the information needed to verify whether the device is properly installed and can operate correctly and safely, plus details of the nature and frequency of the maintenance and calibration needed to ensure that the devices operate properly and safely at all times;
e. Where appropriate, information to avoid certain risks in connection with implantation of the device;
f. Information regarding the risks of reciprocal interference posed by the presence of the device during specific investigations or treatment;
g. The necessary instructions in the event of damage to the sterile packaging and, where appropriate, details of appropriate methods of sterilization;

> *h. If the device is reusable, information on the appropriate processes to allow reuse, including cleaning, disinfection, packaging, and, where appropriate, the method of sterilization of the device to be resterilized, and any restriction on the number of reuses.*
> *Where devices are supplied with the intention that they be sterilized before use, the instructions for cleaning and sterilization must be such that, if correctly followed, the device will still comply with the requirements in Section I.*
> *If the device bears an indication that, the device is for single use, information on known characteristics and technical factors known to the manufacturer that could pose a risk if the device were to be re-used. If in accordance with Section 13.1 no instructions for use are needed, the information must be made available to the user upon request;*
> *i. Details of any further treatment or handling needed before the device can be used (for example, sterilization, final assembly, etc.);*
> *j. In the case of devices emitting radiation for medical purposes, details of the nature, type, intensity, and distribution of this radiation.*
> *The instructions for use must also include details allowing the medical staff to brief the patient on any contra-indications and any precautions to be taken. These details should cover in particular:*
> *k. Precautions to be taken in the event of changes in the performance of the device;*
> *l. Precautions to be taken as regards exposure, in reasonably foreseeable environmental conditions, to magnetic fields, external electrical influences, electrostatic discharge, pressure or variations in pressure, acceleration, thermal ignition sources, etc.;*
> *m. Adequate information regarding the medicinal product or products which the device in question is designed to administer, including any limitations in the choice of substances to be delivered;*
> *n. Precautions to be taken against any special, unusual risks related to the disposal of the device;*
> *o. Medicinal substances, or human blood derivatives incorporated into the device as an integral part in accordance with Section 7.4;*
> *p. Degree of accuracy claimed for devices with a measuring function;*
> *q. Date of issue or the latest revision of the instructions f*or use.

Compliance to the ER is usually shown by completion of a check-list like documentation. Manufacturers are expected to indicate which requirements are applicable and which are not applicable. In addition, they are also required to show which standards they are in compliance with as well as what supporting evidence, they have to show this.

A simple template for manufacturers to use when completing their ER is included in the appendix at the end of this volume.

3.4. CLASSIFICATION RULES UNDER THE MDD

The MDD groups all medical devices into one of four classes, namely Class I, IIa, IIb, and III. Each class is of increasing risk to the patient with medical devices being placed in each respective class based on their safety, function, and intended purpose. Class I devices are those that pose a low risk to the patient and, except for sterile products or measuring devices, can be self-certified by the manufacturer. These devices do not come into contact or interact with the body. Class IIa devices are of a medium risk and are invasive in their interaction with the human body. The methods on invasion however are limited to natural body orifices. Class IIb devices are of a medium risk and are either partially or totally implantable within the human body, and may modify the biological or chemical composition of body fluids. Class III devices are of high risk and require design/clinical trial reviews, product certification and an assessed quality system. These devices affect the functioning of vital organs and/or life-support systems (Figure 3.1).

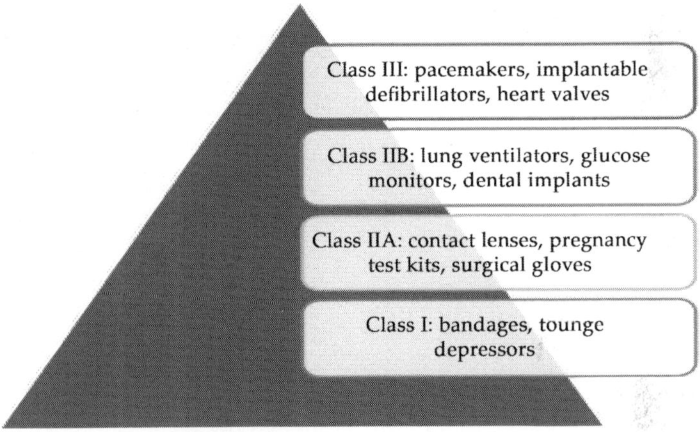

Figure 3.1. Overview of the different medical device classes.

The level of control applied to a medical device is often proportionate to the level of risk it possesses in order to ensure protection of patient health. Medical device classification may also be affected by the time period in which the device performs its intended function. Three definitions for duration of use apply to the directive: *transient* (normally intended for continuous use of less than 60 minutes), *short-term* (normally intended for continuous use of 30 days or less) and *long-term* (normally intended for continuous use of more than 30 days).

Article 9 of the MDD 93/42/EEC: Classification
1. Devices shall be divided into Classes I, IIa, IIb, and III. Classification shall be carried out in accordance with Annex IX. *2. In the event of a dispute between the manufacturer and the notified body concerned, resulting from the application of the classification rules, the matter shall be referred for decision to the competent authority to which the notified body is subject.* *3. Where a Member State considers that the classification rules set out in Annex IX require adaptation in the light of technical progress and any information which becomes available under the information system provided for in Article 10, it may submit a duly substantiated request to the Commission and ask it to take the necessary measures for adaptation of classification rules. The measures designed to amend non-essential elements of this Directive relating to adaptation of classification rules shall be adopted in accordance with the regulatory procedure with scrutiny referred to in Article 7(3).*

Now let's look at the requirements of Annex IX (Classification criteria) in greater detail.

Annex IX of the MDD 93/42/EEC: Classification criteria

I. Definitions

1. Definitions for the classification rules.

1.1. Duration

Transient

Normally intended for continuous use for less than 60 minutes.

Short term

Normally intended for continuous use for not more than 30 days.

Long term

Normally intended for continuous use for more than 30 days.

1.2. Invasive devices

Invasive device

A device which, in whole or in part, penetrates inside the body, either through a body orifice or through the surface of the body.

Body orifice

Any natural opening in the body, as well as the external surface of the eyeball, or any permanent artificial opening, such as a stoma.

Surgically invasive device

An invasive device which penetrates inside the body through the surface of the body, with the aid or in the context of a surgical operation.

For the purposes of this Directive, devices other than those referred to in the previous subparagraph and which produce penetration other than through an established body orifice, shall be treated as surgically invasive devices.

Implantable device

Any device which is intended:

- *To be totally introduced into the human body or;*
- *To replace an epithelial surface or the surface of the eye, by surgical intervention which is intended to remain in place after the procedure.*

Any device intended to be partially introduced into the human body through surgical intervention and intended to remain in place after the procedure for at least 30 days is also considered an implantable device.

1.3. Reusable surgical instrument

Instrument intended for surgical use by cutting, drilling, sawing, scratching, scraping, clamping, retracting, clipping or similar procedures, without connection to any active medical device and which can be reused after appropriate procedures have been carried out.

1.4. Active medical device

Any medical device operation of which depends on a source of electrical energy or any source of power other than that directly generated by the human body or gravity and which acts by converting this energy. Medical devices intended to transmit energy, substances or other elements between an active medical device and the patient, without any significant change, are not considered to be active medical devices. Stand alone software is considered to be an active medical device.

1.5. Active therapeutical device

Any active medical device, whether used alone or in combination with other medical devices, to support, modify, replace or restore biological functions or structures with a view to treatment or alleviation of an illness, injury or handicap.

1.6. Active device for diagnosis

Any active medical device, whether used alone or in combination with other medical devices, to supply information for detecting, diagnosing, monitoring or treating physiological conditions, states of health, illnesses or congenital deformities.

1.7. Central circulatory system

For the purposes of this Directive, 'central circulatory system' means the following vessels:

arteriae pulmonales, aorta ascendens, arcus aorta, aorta descendens to thebifurcatio aortae, arteriae coronariae, arteria carotiscommunis, arteriacarotis externa, arteria carotisinterna, arteriae cerebrales, truncus brachiocephalicus, venae cordis, venae pulmonales, vena cava superior, vena cavainferior.

1.8. Central nervous system

For the purposes of this Directive, 'central nervous system' means brain, meninges, and spinal cord.

II. Implementing Rules

2. Implementing rules

2.1. Application of the classification rules shall be governed by the intended purpose of the devices.

2.2. If the device is intended to be used in combination with another device, the classification rules shall apply separately to each of the devices. Accessories are classified in their own right separately from the device with which they are used.

2.3. Software, which drives a device or influences the use of a device, falls automatically in the same class.

2.4. If the device is not intended to be used solely or principally in a specific part of the body, it must be considered and classified on the basis of the most critical specified use.

2.5. If several rules apply to the same device, based on the performance specified for the device by the manufacturer, the strictest rules resulting in the higher classification shall apply.

2.6. In calculating the duration referred to in Section 1.1 of Chapter I, continuous use means 'an uninterrupted actual use of the device for the intended purpose.' However, where usage of a device is discontinued in order for the device to be replaced immediately by the same or an identical device this shall be considered an extension of the continuous use of the device.

III. Classification

1. Non-invasive devices

1.1. Rule 1

All non-invasive devices are in Class I, unless one of the rules set out hereinafter applies.

1.2. Rule 2

All non-invasive devices intended for channeling or storing blood, body liquids or tissues, liquids or gases for the purpose of eventual infusion, administration or introduction into the body are in Class IIa:

- *If they may be connected to an active medical device in Class IIa or a higher class.*

· *If they are intended for use for storing or channeling blood or other body liquids or for storing organs, parts of organs or body tissues, in all other cases they are in Class I.*

1.3. Rule 3

All non-invasive devices intended for modifying the biological or chemical composition of blood, other body liquids or other liquids intended for infusion into the body are in Class IIb, unless the treatment consists of filtration, centrifugation or exchanges of gas, heat, in which case they are in Class IIa.

1.4. Rule 4

All non-invasive devices which come into contact with injured skin:

· *Are in Class I if they are intended to be used as a mechanical barrier, for compression or for absorption of exudates.*

· *Are in Class IIb if they are intended to be used principally with wounds which have breached the dermis and can only heal by secondary intent.*

· *Are in Class IIa in all other cases, including devices principally intended to manage the micro-environment of a wound.*

2. Invasive devices

2.1. Rule 5

All invasive devices with respect to body orifices, other than surgically invasive devices and which are not intended for connection to an active medical device or which are intended for connection to an active medical device in Class I:

· *Are in Class I if they are intended for transient use.*

· *Are in Class IIa if they are intended for short-term use, except if they are used in the oral cavity as far as the pharynx, in an ear canal up to the ear drum or in a nasal cavity, in which case they are in Class I.*

· *Are in Class IIb if they are intended for long-term use, except if they are used in the oral cavity as far as the pharynx, in an ear canal up to the ear drum or in a nasal cavity and are not liable to be absorbed by the mucous membrane, in which case they are in Class IIa.*

All invasive devices with respect to body orifices, other than surgically invasive devices, intended for connection to an active medical device in Class IIa or a higher class, are in Class IIa.

2.2. Rule 6

All surgically invasive devices intended for transient use are in Class IIa unless they are:

- *Intended specifically to control, diagnose, monitor or correct a defect of the heart or of the central circulatory system through direct contact with these parts of the body, in which case they are in Class III.*
- *Reusable surgical instruments, in which case they are in Class I.*
- *Intended specifically for use in direct contact with the central nervous system, in which case they are in Class III.*
- *Intended to supply energy in the form of ionizing radiation in which case they are in Class IIb.*
- *Intended to have a biological effect or to be wholly or mainly absorbed in which case they are in Class IIb.*
- *Intended to administer medicines by means of a delivery system, if this is done in a manner that is potentially hazardous taking account of the mode of application, in which case they are in Class IIb.*

2.3. Rule 7

All surgically invasive devices intended for short-term use are in Class IIa unless they are intended:

- *Either specifically to control, diagnose, monitor or correct a defect of the heart or of the central circulatory system through direct contact with these parts of the body, in which case they are in Class III.*
- *Or specifically for use in direct contact with the central nervous system, in which case they are in Class III.*
- *Or to supply energy in the form of ionizing radiation in which case they are in Class IIb.*
- *Or to have a biological effect or to be wholly or mainly absorbed in which case they are in Class III.*

- *Or to undergo chemical change in the body, except if the devices are placed in the teeth, or to administer medicines, in which case they are in Class IIb.*

2.4. Rule 8

All implantable devices and long-term surgically invasive devices are in Class IIb unless they are intended:

- *To be placed in the teeth, in which case they are in Class IIa.*
- *To be used in direct contact with the heart, the central circulatory system or the central nervous system, in which case they are in Class III.*
- *To have a biological effect or to be wholly or mainly absorbed, in which case they are in Class III.*
- *Or to undergo chemical change in the body, except if the devices are placed in the teeth, or to administer medicines, in which case they are in Class III.*

3. Additional rules applicable to active devices

3.1. Rule 9

All active therapeutic devices intended to administer or exchange energy are in Class IIa unless their characteristics are such that they may administer or exchange energy to or from the human body in a potentially hazardous way, taking account of the nature, the density and site of application of the energy, in which case they are in Class IIb. All active devices intended to control or monitor the performance of active therapeutic devices in Class IIb, or intended directly to influence the performance of such devices are in Class IIb.

3.2. Rule 10

Active devices intended for diagnosis are in Class IIa:

- *If they are intended to supply energy which will be absorbed by the human body, except for devices used to illuminate the patient's body, in the visible spectrum.*
- *If they are intended to image in vivo distribution of radiopharmaceuticals.*

- *If they are intended to allow direct diagnosis or monitoring of vital physiological processes, unless they are specifically intended for monitoring of vital physiological parameters, where the nature of variations is such that it could result in immediate danger to the patient, for instance variations in cardiac performance, respiration, activity of CNS in which case they are in Class IIb.*

Active devices intended to emit ionizing radiation and intended for diagnostic and therapeutic interventional radiology including devices which control or monitor such devices, or which directly influence their performance, are in Class IIb.

3.3 Rule 11

All active devices intended to administer and/or remove medicines, body liquids or other substances to or from the body are in Class IIa, unless this is done in a manner:

- *That is potentially hazardous, taking account of the nature of the substances involved, of the part of the body concerned and of the mode of application in which case they are in Class IIb.*

3.4. Rule 12

All other active devices are in Class I.

4. Special Rules

4.1. Rule 13

All devices incorporating, as an integral part, a substance which, if used separately, can be considered to be a medicinal product, as defined in Article 1 of Directive 2001/83/EC, and which is liable to act on the human body with action ancillary to that of the devices, are in Class III.

All devices incorporating, as an integral part, a human blood derivative are in Class III.

4.2. Rule 14

All devices used for contraception or the prevention of the transmission of sexually transmitted diseases are in Class IIb, unless they are implantable or long term invasive devices, in which case they are in Class III.

4.3. Rule 15

All devices intended specifically to be used for disinfecting, cleaning, rinsing, or, when appropriate, hydrating contact lenses are in Class IIb.

All devices intended specifically to be used for disinfecting medical devices are in Class IIa. Unless they are specifically to be used for disinfecting invasive devices in which case, they are in Class IIb.

This rule does not apply to products that are intended to clean medical devices other than contact lenses by means of physical action.

4.4. Rule 16

Devices specifically intended for recording of X-ray diagnostic images are in Class IIa.

4.5. Rule 17

All devices manufactured utilizing animal tissues or derivatives rendered on-viable are Class III except where such devices are intended to come into contact with intact skin only.

5. Rule 18

*By derogation from other rules, blood bags are in Class II*b.

3.5. CONFORMITY ASSESSMENT ROUTES AS PER THE MDD

Within the MDD, manufacturers are given a choice of different paths which they can follow to show they meet the MDD requirements. It's important to remember that the class of product will determine the overall route that must be followed. Conformity procedures address two main stages of the product life cycle: 1) design and 2) manufacture.

With regards to design, manufacturers must be able to provide evidence showing how the device meets the ER. This evidence should be held in a product technical file or design dossier. For manufacture, a documented quality system must be in place to ensure that the devices continue to comply with the ER and that they are consistent with the information in the technical file.

Routes of conformity assessment are identified in annexes II, III, IV, V, VI, VII, and VIII of the MDD. Many of these annexes contain requirements that the manufacturer's quality system be assessed and are in compliance with applicable standards. As classification of the device determines the appropriate conformity assessment route, it is critical that this step be done by individuals who understand the regulation. Often manufacturers seek advice from a notified body or external consultant to assist them with carrying out their conformity assessment.

An overview of the different conformity assessment routes as provided in the MDD are provided below:

Article 11 of the MDD 93/42/EEC: Conformity assessment procedures
1. In the case of devices falling within Class III, other than devices which are custom-made or intended for clinical investigations, the manufacturer shall, in order to affix the CE marking, either: *a. Follow the procedure relating to the EC declaration of conformity set out in Annex II (full quality assurance); or* *b. Follow the procedure relating to the EC type-examination set out in Annex III, coupled with:* *i. The procedure relating to the EC verification set out in Annex IV; or* *ii. The procedure relating to the EC declaration of conformity set out in Annex V (production quality assurance).* *2. In the case of devices falling within Class IIa, other than devices which are custom-made or intended for clinical investigations, the manufacturer shall, in order to affix the CE marking, follow the procedure relating to the EC declaration of conformity set out in Annex VII, coupled with either:* *a. The procedure relating to the EC verification set out in Annex IV; or* *b. The procedure relating to the EC declaration of conformity set out in Annex V (production quality assurance); or* *c. The procedure relating to the EC declaration of conformity set out in Annex VI (product quality assurance).* *Instead of applying these procedures, the manufacturer may also follow the procedure referred to in paragraph 3(a).*

3. In the case of devices falling within Class IIb, other than devices which are custom-made or intended for clinical investigations, the manufacturer shall, in order to affix the CE marking, either:
a. Follow the procedure relating to the EC declaration of conformity set out in Annex II (full quality assurance); in this case, point 4 of Annex II is not applicable; or
b. Follow the procedure relating to the EC type-examination set out in Annex III, coupled with:
i. The procedure relating to the EC verification set out in Annex IV; or
ii. The procedure relating to the EC declaration of conformity set out in Annex V (production quality assurance); or
iii. The procedure relating to the EC declaration of conformity set out in Annex VI (product quality assurance).
4. The Commission shall, no later than five years from the date of implementation of this Directive, submit a report to the Council on the operation of the provisions referred to in Article 10(1), Article 15(1), in particular in respect of Class I and Class IIa devices, and on the operation of the provisions referred to in Annex II, Section 4.3 second and third subparagraphs and in Annex III, Section 5 second and third subparagraphs to this Directive, accompanied, if necessary, by appropriate proposals.
5. In the case of devices falling within Class I, other than devices which are custom-made or intended for clinical investigations, the manufacturer shall, in order to affix the CE marking, follow the procedure referred to in Annex VII and draw up the EC declaration of conformity required before placing the device on the market.
6. In the case of custom-made devices, the manufacturer shall follow the procedure referred to in Annex VIII and draw up the statement set out in that Annex before placing each device on the market. Member States may require that the manufacturer shall submit to the competent authority a list of such devices which have been put into service in their territory.

7. During the conformity assessment procedure for a device, the manufacturer, and/or the notified body shall take account of the results of any assessment and verification operations which, where appropriate, have been carried out in accordance with this Directive at an intermediate stage of manufacture.
8. The manufacturer may instruct his authorized representative to initiate the procedures provided for in Annexes III, IV, VII, and VIII.
9. Where the conformity assessment procedure involves the intervention of a notified body, the manufacturer, or his authorized representative, may apply to a body of his choice within the framework of the tasks for which the body has been notified.
10. The notified body may require, where duly justified, any information or data, which is necessary for establishing and maintaining the attestation of conformity in view of the chosen procedure.
11. Decisions taken by the NBs in accordance with Annexes II, III, V, and VI shall be valid for a maximum of five years and may be extended on application, made at a time agreed in the contract signed by both parties, for further periods of a maximum length of five years.
12. The records and correspondence relating to the procedures referred to in paragraphs 1 to 6 shall be in an official language of the Member State in which the procedures are carried out and/or in another Community language acceptable to the notified body.
13. By derogation from paragraphs 1 to 6, the CA may authorize, on duly justified request, the placing on the market and putting into service, within the territory of the Member State concerned, of individual devices for which the procedures referred to in paragraphs 1 to 6 have not been carried out and the use of which is in the interest of protection of health.
14. The measures designed to amend non-essential elements of this Directive, by supplementing it, relating to the means by which, in the light of technical progress and considering the intended users of the devices concerned, the information laid down in Annex I Section 13.1 may be set out, shall be adopted in accordance with the regulatory procedure with scrutiny referred to in Article 7(3).

The Medical Devices Directive (MDD) applies to all general medical devices that are not covered by the Active Implantable Medical Devices Directive or the *In vitro* Diagnostic Medical Devices Directive. Medical devices must be correctly classified into one of the defined different classes, based on risk and intended use, which then determines the relevant conformity assessment procedure that should be followed. For products classified with medium to high degree of risk (class Is, Im, IIa, IIb, and III) the MDD requires a conformity assessment procedure involving a notified body.

3.6. DETERMINING THE BEST COMPLIANCE ROUTE FOR YOUR DEVICE

The compliance routes available to be able to CE mark your device, in accordance with the Medical Devices Directive will depend on the classification of the device involved. It is not always practical or even feasible to apply the most rigorous conformity assessment procedure to a device. Instead, manufacturers should opt for a graduated level of control which categorizes devices based on the level of inherent risk or potential hazard. For this reason, the correct classification of a device is therefore critical before a manufacturer can commence with a compliance assessment. In depth, guidance on performing the correct classification of a device can be found in the European Commission Guidance Document: MEDDEV 2.4/1-Classification of Medical Devices.

A summary of the conformity assessment routes under the MDD are provided in Table 3.1.

Table 3.1. Conformity Assessment Routes under the MDD

Risk Class of the Device	Overview of CE Marking Process
Class I	Class I devices follow a self-declaration of conformity route, unless the device is sold as sterile (Class Is) or has a measurement function (Class Im.) In these cases, the involvement of a notified body is required.
Class IIa and IIb	Class IIa and IIb devices require the services of a notified body to approve the declaration of conformity through a conformity assessment.
Class III	Class III devices represent the highest risk and are evaluated by a specialist team of assessors within the notified body.

While several different conformity assessment routes are available to the manufacturer, the most common approach is Annex II. Once you have been issued with a certificate by the Notified Body, you may affix CE marking to your device and place it on the market. Most member states also require you to register your device with the relevant national authorities (Table 3.2).

Table 3.2. Overview of Conformity Assessment Routes as Specified by the MDD

Annex II of the MDD 93/42/EEC: EC Declaration of Conformity (Full quality assurance system)

1. The manufacturer must ensure application of the quality system approved for the design, manufacture, and final inspection of the products concerned, as specified in Section 3 and is subject to audit as laid down in Sections 3.3 and 4 and to Community surveillance as specified in Section 5.

2. The EC declaration of conformity is the procedure whereby the manufacturer who fulfills the obligations imposed by Section 1 ensures and declares that the products concerned meet the provisions of this Directive which apply to them.

The manufacturer must affix the CE marking in accordance with Article 17 and draw up a written declaration of conformity. This declaration must cover one or more medical devices manufactured, clearly identified by means of product name, product code, or other unambiguous reference, and must be kept by the manufacturer.

3. Quality system.

3.1. The manufacturer must lodge an application for assessment of his quality system with a notified body.

The application must include:

- *The name and address of the manufacturer and any additional manufacturing site covered by the quality system.*
- *All the relevant information on the product or product category covered by the procedure.*
- *A written declaration that no application has been lodged with any other notified body for the same product-related quality system.*
- *The documentation on the quality system.*
- *An undertaking by the manufacturer to fulfill the obligations imposed by the quality system approved.*
- *An undertaking by the manufacturer to keep the approved quality system adequate and efficacious.*
- *An undertaking by the manufacturer to institute and keep up to date a systematic procedure to review experience gained from devices in the post-production phase, including the provisions referred to in Annex X, and to implement appropriate means to apply any necessary corrective action.*

- *This undertaking must include an obligation for the manufacturer to notify the CA of the following incidents immediately on learning of them:*

i. Any malfunction or deterioration in the characteristics and/or performance of a device, as well as any inadequacy in the instructions for use which might lead to or might have led to the death of a patient or user or to a serious deterioration in his state of health;

ii. Any technical or medical reason connected with the characteristics or performance of a device leading for the reasons referred to in subparagraph (i) to systematic recall of devices of the same type by the manufacturer.

3.2. Application of the quality system must ensure that the products conform to the provisions of this Directive which apply to them at every stage, from design to final inspection. All the elements, requirements, and provisions adopted by the manufacturer for his quality system must be documented in a systematic and orderly manner in the form of written policies and procedures such as quality programmes, quality plans, quality manuals, and quality records.

It shall include in particular the corresponding documentation, data, and records arising from the procedures referred to in point (c).

It shall include in particular an adequate description of:

a. The manufacturer's quality objectives;

b. The organization of the business and in particular:

- *The organizational structures, the responsibilities of the managerial staff and their organizational authority where quality of design and manufacture of the products is concerned.*
- *The methods of monitoring the efficient operation of the quality system and in particular its ability to achieve the desired quality of design and of product, including control of products which fail to conform.*
- *Where the design, manufacture, and/or final inspection and testing of the products, or elements thereof, is carried out by a third party, the methods of monitoring the efficient operation of the quality system and in particular the type and extent of control applied to the third-party;*

c The procedures for monitoring and verifying the design of the products, including the corresponding documentation, and in particular:

- *A general description of the product, including any variants planned, and its intended use(s).*
- *The design specifications, including the standards which will be applied and the results of the risk analysis, and also a description of the solutions adopted to fulfill the ER which apply to the products if the standards referred to in Article 5 are not applied in full.*
- *The techniques used to control and verify the design and the processes and systematic measures which will be used when the products are being designed.*

- *If the device is to be connected to other device(s) in order to operate as intended, proof must be provided that it conforms to the ER when connected to any such device(s)having the characteristics specified by the manufacturer.*
- *A statement indicating whether or not the device incorporates, as an integral part, a substance or a human blood derivative referred to in section 7.4 of Annex I and the data on the tests conducted in this connection required to assess the safety, quality, and usefulness of that substance or human blood derivative, taking account of the intended purpose of the device.*
- *A statement indicating whether or not the device is manufactured utilizing tissues of animal origin as referred to in Commission Directive 2003/32/EC (1).*
- *The solutions adopted as referred to in Annex I, Chapter I, Section 2.*
- *The pre-clinical evaluation.*
- *The clinical evaluation referred to in Annex X.*
- *The draft label and, where appropriate, instructions for use.*

d. The inspection and quality assurance techniques at the manufacturing stage and in particular:

- *The processes and procedures which will be used, particularly as regards sterilization, purchasing, and the relevant documents.*
- *The product identification procedures drawn up and kept up to date from drawings, specifications or other relevant documents at every stage of manufacture;*

e. The appropriate tests and trials which will be carried out before, during and after manufacture, the frequency with which they will take place, and the test equipment used; it must be possible to trace back the calibration of the test equipment adequately.

3.3. The notified body must audit the quality system to determine whether it meets the requirements referred to in Section 3.2. It must presume that quality systems which implement the relevant harmonized standards conform to these requirements.

The assessment team must include at least one member with past experience of assessments of the technology concerned. The assessment procedure must include an assessment, on a representative basis, of the documentation of the design of the product(s) concerned, an inspection on the manufacturer's premises and, in duly substantiated cases, on the premises of the manufacturer's suppliers and/or subcontractors to inspect the manufacturing processes.

The decision is notified to the manufacturer. It must contain the conclusions of the inspection and a reasoned assessment.

3.4. The manufacturer must inform the notified body which approved the quality system of any plan for substantial changes to the quality system or the product-range covered. The notified body must assess the changes proposed and verify whether after these changes the quality system still meets the requirements referred to in Section 3.2. It must notify the manufacturer of its decision. This decision must contain the conclusions of the inspection and a reasoned assessment.

4. Examination of the design of the product.

4.1. In addition to the obligations imposed by Section 3, the manufacturer must lodge with the notified body an application for examination of the design dossier relating to the product which he plans to manufacture and which falls into the category referred to in Section 3.1.

4.2. The application must describe the design, manufacture, and performances of the product in question. It must include the documents needed to assess whether the product conforms to the requirements of this Directive, as referred to in Section 3.2 (c).

4.3. The notified body must examine the application and, if the product conforms to the relevant provisions of this Directive, issue the application with an EC design-examination certificate. The notified body may require the application to be completed by further tests or proof to allow assessment of conformity with the requirements of the Directive. The certificate must contain the conclusions of the examination, the conditions of validity, the data needed for identification of the approved design, where appropriate, and a description of the intended purpose of the product.

In the case of devices referred to in Annex I, Section 7.4, second paragraph, the notified body shall, as regards the aspects referred to in that section, consult one of the CA designated by the Member States in accordance with Directive 2001/83/EC or the EMEA before taking a decision. The opinion of the competent national authority or the EMEA must be drawn up within 210 days after receipt of valid documentation. The scientific opinion of the competent national authority or the EMEA must be included in the documentation concerning the device. The notified body will give due consideration to the views expressed in this consultation when making its decision. It will convey its final decision to the competent body concerned.

In the case of devices referred to in Annex I, Section 7.4, third paragraph, the scientific opinion of the EMEA must be included in the documentation concerning the device. The opinion of the EMEA must be drawn up within 210 days after receipt of valid documentation. The notified body will give due consideration to the opinion of the EMEA when making its decision. The notified body may not deliver the certificate if the EMEA's scientific opinion is unfavorable. It will convey its final decision to the EMEA.

In the case of devices manufactured utilizing tissues of animal origin as referred to in Directive 2003/32/EC, the notified body must follow the procedures referred to in that Directive.

4.4. Changes to the approved design must receive further approval from the notified body which issued the EC design-examination certificate wherever the changes could affect conformity with the ER of the Directive or with the conditions prescribed for use of the product. The applicant shall inform the notified body which issued the EC design-examination certificate of any such changes made to the approved design. This additional approval must take the form of a supplement to the EC design examination certificate.

5. Surveillance.

5.1. The aim of surveillance is to ensure that the manufacturer duly fulfills the obligations imposed by the approved quality system.

5.2. The manufacturer must authorize the notified body to carry out all the necessary inspections and supply it with all relevant information, in particular:

- *The documentation on the quality system.*
- *The data stipulated in the part of the quality system relating to design, such as the results of analyses, calculations, tests, the solutions adopted as referred to in Annex I, Chapter I, Section 2, pre-clinical, and clinical evaluation, post-market clinical follow-up plan and the results of the post-market clinical follow-up, if applicable, etc.*
- *The data stipulated in the part of the quality system relating to manufacture, such as inspection reports and test data, calibration data, qualification reports of the personnel concerned, etc.*

5.3. The notified body must periodically carry out appropriate inspections and assessments to make sure that the manufacturer applies the approved quality system and must supply the manufacturer with an assessment report.

5.4. In addition, the notified body may pay unannounced visits to the manufacturer. At the time of such visits, the notified body may, where necessary, carry out or ask for tests in order to check that the quality system is working properly. It must provide the manufacturer with an inspection report and, if a test has been carried out, with a test report.

6. Administrative provisions.

6.1. The manufacturer or his authorized representative must, for a period ending at least five years, and in the case of implantable devices at least 15 years, after the last product has been manufactured, keep at the disposal of the national authorities:

- *The declaration of conformity.*
- *The documentation referred to in the fourth indent of Section 3.1 and in particular the documentation, data, and records referred to in the second paragraph of Section 3.2.*
- *The changes referred to in Section 3.4.*
- *The documentation referred to in Section 4.2, and*
- *The decisions and reports from the notified body as referred to in Sections 3.3, 4.3, 4.4, 5.3, and 5.4.*

7. Application to devices in Classes IIa and IIb.
7.1. In line with Article 11(2) and (3), this Annex may apply to products in Classes IIa and IIb. Section 4, however, does not apply.
7.2. For devices in Class IIa the notified body shall assess, as part of the assessment in Section 3.3, the technical documentation as described in Section 3.2(c) for at least one representative sample for each device subcategory for compliance with the provisions of this Directive.
7.3. For devices in Class IIb the notified body shall assess, as part of the assessment in Section 3.3, the technical documentation as described in Section 3.2(c) for at least one representative sample for each generic device group for compliance with the provisions of this Directive.
7.4. In choosing representative sample(s) the notified body shall take into account the novelty of the technology, similarities in design, technology, manufacturing, and sterilization methods, the intended use and the results of any previous relevant assessments (e.g., with regard to physical, chemical or biological properties) that have been carried out in accordance with this Directive. The notified body shall document and keep available to the competent authority its rationale for the sample(s) taken.
7.5. Further samples shall be assessed by the notified body as part of the surveillance assessment referred to in Section 5.
8. Application to the devices referred to Article 1(4a).
Upon completing the manufacture of each batch of devices referred to in Article 1(4a), the manufacturer shall inform the notified body of the release of the batch of devices and send to it the official certificate concerning the release of the batch of human blood derivative used in the device, issued by a State laboratory or a laboratory designated for that purpose by a Member State in accordance with Article 114(2) of Directive 2001/83/EC.

Annex III of the MDD 93/42/EEC: EC Type-Examination

1. EC type-examination is the procedure whereby a notified body ascertains and certifies that a representative sample of the production covered fulfills the relevant provisions of this Directive.
2. The application includes:

- *The name and address of the manufacturer and the name and address of the authorized representative if the application is lodged by the representative.*
- *The documentation described in Section 3 needed to assess the conformity of the representative sample of the production in question, hereinafter referred to as the 'type,' with the requirements of this Directive. The applicant must make a 'type' available to the notified body. The notified body may request other samples as necessary.*

- *A written declaration that no application has been lodged with any other notified body for the same type.*

3. The documentation must allow an understanding of the design, the manufacture and the performances of the product and must contain the following items in particular:

- *A general description of the type, including any variants planned, and its intended use(s).*
- *Design drawings, methods of manufacture envisaged, in particular as regards sterilization, and diagrams of components, sub-assemblies, circuits, etc.*
- *The descriptions and explanations necessary to understand the above-mentioned drawings and diagrams and the operation of the product.*
- *A list of the standards referred to in Article 5, applied in full or in part, and descriptions of the solutions adopted to meet the ER if the standards referred to in Article 5 have not been applied in full.*
- *The results of the design calculations, risk analysis, investigations, technical tests, etc. carried out.*
- *A statement indicating whether or not the device incorporates, as an integral part, a substance, or human blood derivative, referred to in Section 7.4 of Annex I, and the data on the tests conducted in this connection which are required to assess the safety, quality, and usefulness of that substance, or human blood derivative, taking account of the intended purpose of the device.*
- *A statement indicating whether or not the device is manufactured utilizing tissues of animal origin as referred to in Directive 2003/32/EC.*
- *The solutions adopted as referred to in Annex I, Chapter I, Section 2.*
- *The pre-clinical evaluation.*
- *The clinical evaluation referred to in Annex X.*
- *The draft label and, where appropriate, instructions for use.*

4. The notified body must:

4.1. Examine and assess the documentation and verify that the type has been manufactured in conformity with that documentation; it must also record the items designed in conformity with the applicable provisions of the standards referred to in Article 5, as well as the items not designed on the basis of the relevant provisions of the above-mentioned standards;

4.2. Carry out or arrange for the appropriate inspections and the tests necessary to verify whether the solutions adopted by the manufacturer meet the ER of this Directive if the standards referred to in Article 5 have not been applied; if the device is to be connected to other device(s) in order to operate as intended, proof must be provided that it conforms to the ER when connected to any such device(s) having the characteristics specified by the manufacturer;

4.3. Carry out or arrange for the appropriate inspections and the tests necessary to verify whether, if the manufacturer has chosen to apply the relevant standards, these have actually been applied;

4.4. Agree with the applicant on the place where the necessary inspections and tests will be carried out.

5. If the type conforms to the provisions of this Directive, the notified body issues the applicant with an EC type-examination certificate. The certificate must contain the name and address of the manufacturer, the conclusions of the inspection, the conditions of validity, and the data needed for identification of the type approved. The relevant parts of the documentation must be annexed to the certificate and a copy kept by the notified body.

In the case of devices referred to in Annex I, Section 7.4, second paragraph, the notified body shall, as regards the aspects referred to in that section, consult one of the authorities designated by the Member States in accordance with Directive 2001/83/EC or the EMEA before taking a decision. The opinion of the competent national authority or the EMEA must be drawn up within 210 days after receipt of valid documentation. The scientific opinion of the competent national authority or the EMEA must be included in the documentation concerning the device. The notified body will give due consideration to the views expressed in this consultation when making its decision. It will convey its final decision to the competent body concerned.

In the case of devices referred to in Annex I, Section 7.4, third paragraph, the scientific opinion of the EMEA must be included in the documentation concerning the device. The opinion of the EMEA must be drawn up within 210 days after receipt of valid documentation. The notified body will give due consideration to the opinion of the EMEA when making its decision. The notified body may not deliver the certificate if the EMEA's scientific opinion is unfavorable. It will convey its final decision to the EMEA.

In the case of devices manufactured utilizing tissues of animal origin as referred to in Directive 2003/32/EC, the notified body must follow the procedures referred to in that Directive.

6. The applicant must inform the notified body which issued the EC type examination certificate of any significant change made to the approved product.

Changes to the approved product must receive further approval from the notified body which issued the EC type-examination certificate wherever the changes may affect conformity with the ER or with the conditions prescribed for use of the product. This new approval must, where appropriate, take the form of a supplement to the initial EC type examination certificate.

7. Administrative provisions.

7.2. Other NBs may obtain a copy of the EC type-examination certificates and/or the supplements thereto. The Annexes to the certificates must be made available to other NBs on reasoned application, after the manufacturer has been informed.

7.3. The manufacturer or his authorized representative must keep with the technical documentation copies of EC type-examination certificates and their additions for a period ending at least five years after the last device has been manufactured. In the case of implantable devices, the period shall be at least 15 years after the last product has been manufactured.

Annex IV of the MDD 93/42/EEC: EC Verification
1. EC verification is the procedure whereby the manufacturer or his authorized representative ensures and declares that the products which have been subject to the procedure set out in Section 4 conform to the type described in the EC type-examination certificate and meet the requirements of this Directive which apply to them. *2. The manufacturer must take all the measures necessary to ensure that the manufacturing process produces products which conform to the type described in the EC type-examination certificate and to the requirements of the Directive which apply to them. Before the start of manufacture, the manufacturer must prepare documents defining the manufacturing process, in particular as regards sterilization where necessary, together with all the routine, pre-established provisions to be implemented to ensure homogeneous production and, where appropriate, conformity of the products with the type described in the EC type-examination certificate and with the requirements of this Directive which apply to them. The manufacturer must affix the CE marking in accordance with Article 17 and draw up declaration of conformity.* *In addition, for products placed on the market in sterile condition, and only for those aspects of the manufacturing process designed to secure and maintain sterility, the manufacturer must apply the provisions of Annex V, Sections 3 and 4.* *3. The manufacturer must undertake to institute and keep up to date a systematic procedure to review experience gained from devices in the post-production phase, including the provisions referred to in Annex X, and to implement appropriate means to apply any necessary corrective action. This undertaking must include an obligation for the manufacturer to notify the CA of the following incidents immediately on learning of them:* *i. Any malfunction or deterioration in the characteristics and/or performance of a device, as well as any inadequacy in the labeling or the instructions for use which might lead to or might have led to the death of a patient or user or to a serious deterioration in his state of health;* *ii. Any technical or medical reason connected with the characteristics or performance of a device for the reasons referred to in subparagraph (i) leading to systematic recall of devices of the same type by the manufacturer.* *4. The notified body must carry out the appropriate examinations and tests in order to verify the conformity of the product with the requirements of the Directive either by examining and testing every product as specified in Section 5 or by examining and testing products on a statistical basis as specified in Section 6, as the manufacturer decides.* *The aforementioned checks do not apply to those aspects of the manufacturing process designed to secure sterility.* *5. Verification by examination and testing of every product.*

5.1. Every product is examined individually and the appropriate tests defined in the relevant standard(s) referred to in Article 5 or equivalent tests must be carried out in order to verify, where appropriate, the conformity of the products with the EC type described in the type-examination certificate and with the requirements of the Directive which apply to them.

5.2. The notified body must affix, or have affixed its identification number to each approved product and must draw up a written certificate of conformity relating to the tests carried out.

6. Statistical verification.

6.1. The manufacturer must present the manufactured products in the form of homogeneous batches.

6.2. A random sample is taken from each batch. The products which make up the sample are examined individually and the appropriate tests defined in the relevant standard(s) referred to in Article 5 or equivalent tests must be carried out to verify, where appropriate, the conformity of the products with the type described in the EC type-examination certificate and with the requirements of the Directive which apply to them in order to determine whether to accept or reject the batch.

6.3. Statistical control of products will be based on attributes and/or variables, entailing sampling schemes with operational characteristics which ensure a high level of safety and performance according to the state of the art. The sampling schemes will be established by the harmonized standards referred to in Article 5, taking account of the specific nature of the product categories in question.

6.4. If the batch is accepted, the notified body affixes or has affixed its identification number to each product and draws up a written certificate of conformity relating to the tests carried out. All products in the batch may be put on the market except any in the sample which failed to conform.

If a batch is rejected, the competent notified body must take appropriate measures to prevent the batch from being placed on the market. In the event of frequent rejection of batches, the notified body may suspend the statistical verification. The manufacturer may, on the responsibility of the notified body, affix the notified body's identification number during the manufacturing process.

7. Administrative provisions.

The manufacturer or his authorized representative must, for a period ending at least five years, and in the case of implantable devices at least 15 years, after the last product has been manufactured, make available to the national authorities:

- *The declaration of conformity.*
- *The documentation referred to in Section 2.*
- *The certificates referred to in Sections 5.2 and 6.4.*
- *Where appropriate, the type-examination certificate referred to in Annex III.*

8. Application to devices in Class IIa.

In line with Article 11 (2), this Annex may apply to products in Class IIa, subject to the following:

8.1. In derogation from Sections 1 and 2, by virtue of the declaration of conformity the manufacturer ensures and declares that the products in Class IIa are manufactured in conformity with the technical documentation referred to in Section 3 of Annex VII and meet the requirements of this Directive which apply to them; *8.2. In derogation from Sections 1, 2, 5 and 6, the verifications conducted by the notified body are intended to confirm the conformity of the products in Class IIa with the technical documentation referred to in Section 3 of Annex VII.* *9. Application to devices referred to in Article 1(4a).* *In the case of section 5, upon completing the manufacture of each batch of devices referred to in Article 1(4a), and in the case of verification under section 6, the manufacturer shall inform the notified body of the release of this batch of devices and send to it the official certificate concerning the release of the batch of human blood derivative used in the device issued by a State laboratory or a laboratory designated for that purpose by a Member State in accordance with Article 114(2) of Directive 2001/83/EC.*
Annex V of the MDD 93/42/EEC: EC Declaration of Conformity (Production quality assurance)
1. The manufacturer must ensure application of the quality system approved for the manufacture of the products concerned and carry out the final inspection, as specified in Section 3, and is subject to the Community surveillance referred to in Section 4. *2. The EC declaration of conformity is the part of the procedure whereby the manufacturer who fulfills the obligations imposed by Section 1 ensures and declares that the products concerned conform to the type described in the EC type-examination certificate and meet the provisions of this Directive which apply to them.* *The manufacturer must affix the CE marking in accordance with Article 17 and draw up a written declaration of conformity. This declaration must cover one or more medical devices manufactured, clearly identified by means of product name, product code or other unambiguous reference, and must be kept by the manufacturer.* *3. Quality system.* *3.1. The manufacturer must lodge an application for assessment of his quality system with a notified body.* *The application must include:* • *The name and address of the manufacturer.* • *All the relevant information on the product or product category covered by the procedure.* • *A written declaration that no application has been lodged with any other notified body for the same products.* • *The documentation on the quality system.* • *An undertaking to fulfill the obligations imposed by the quality system disapproved.*

- *An undertaking to maintain the practicability and effectiveness of the approved quality system.*
- *Where appropriate, the technical documentation on the types approved and a copy of the EC type-examination certificates.*
- *An undertaking by the manufacturer to institute and keep up to date a systematic procedure to review experience gained from devices in the post-production phase, including the provisions referred to in Annex X, and to implement appropriate means to apply any necessary corrective action. This undertaking must include an obligation for the manufacturer to notify the CA of the following incidents immediately on learning of them:*

i. Any malfunction or deterioration in the characteristics and/or performance of a device, as well as any inadequacy in the labeling or the instructions for use which might lead to or might have led to the death of a patient or user or to a serious deterioration in his state of health;

ii. Any technical or medical reason connected with the characteristics or performance of a device for the reasons referred to in subparagraph (i) above leading to a systematic recall of devices of the same type by the manufacturer.

3.2. Application of the quality system must ensure that the products conform to the type described in the EC type-examination certificate.

All the elements, requirements, and provisions adopted by the manufacturer for his quality system must be documented in a systematic and orderly manner in the form of written policy statements and procedures. This quality system documentation must permit uniform interpretation of the quality policy and procedures such as quality programmes, plans, manuals, and records.

It must include in particular an adequate description of:

a. The manufacturer's quality objectives;

b. The organization of the business and in particular:

· The organizational structures, the responsibilities of the managerial staff and their organizational authority where manufacture of the products is concerned.

- *The methods of monitoring the efficient operation of the quality system and in particular its ability to achieve the desired quality of product, including control of products which fail to conform.*
- *Where the manufacture and/or final inspection and testing of the products, or elements thereof, are carried out by a third party, the methods of monitoring the efficient operation of the quality system and in particular the type and extent of control applied to the third-party;*

c. The inspection and quality assurance techniques at the manufacturing stage and in particular:

- *The processes and procedures which will be used, particularly as regards sterilization, purchasing, and the relevant documents.*
- *The product identification procedures drawn up and kept up to date from drawings, specifications or other relevant documents at every stage of manufacture;*

d. The appropriate tests and trials to be carried out before, during, and after manufacture, the frequency with which they will take place, and the test equipment used; it must be possible adequately to trace back the calibration of the test equipment.

3.3. The notified body must audit the quality system to determine whether it meets the requirements referred to in Section 3.2. It must presume that quality systems which implement the relevant harmonized standards conform to these requirements.

The assessment team must include at least one member with past experience of assessments of the technology concerned. The assessment procedure must include an inspection on the manufacturer's premises and, in duly substantiated cases, on the premises of the manufacturer's suppliers to inspect the manufacturing processes.

The decision must be notified to the manufacturer after the final inspection and contain the conclusions of the inspection and a reasoned assessment.

3.4. The manufacturer must inform the notified body which approved the quality system of any plan for substantial changes to the quality system.

The notified body must assess the changes proposed and verify whether after these changes the quality system still meets the requirements referred to in Section 3.2.

After the above-mentioned information has been received, the decision is notified to the manufacturer. It must contain the conclusions of the inspection and a reasoned assessment.

4. Surveillance.

4.1. The aim of surveillance is to ensure that the manufacturer duly fulfills the obligations imposed by the approved quality system.

4.2. The manufacturer authorizes the notified body to carry out all the necessary inspections and must supply it with all relevant information, in particular:

- *The documentation on the quality system.*
- *The technical documentation,—the data stipulated in the part of the quality system relating to manufacture, such as inspection reports and test data, calibration data, qualification reports of the personnel concerned, etc.*

4.3. The notified body must periodically carry out appropriate inspections and assessments to make sure that the manufacturer applies the approved quality system and supply the manufacturer with an assessment report.

4.4. In addition, the notified body may pay unannounced visits to the manufacturer. At the time of such visits, the notified body may, where necessary, carry out or ask for tests in order to check that the quality system is working properly. It must provide the manufacturer with an inspection report and, if a test has been carried out, with a test report.

5. Administrative provisions.
5.1. The manufacturer or his authorized representative must, for a period ending at least five years, and in the case of implantable devices at least 15 years, after the last product has been manufactured, make available to the national authorities:

- *The declaration of conformity.*
- *The documentation referred to in the fourth indent of Section 3.1.*
- *The changes referred to in Section 3.4.*
- *The documentation referred to in the seventh indent of Section 3.1.*
- *The decisions and reports from the notified body as referred to in Sections 4.3 and 4.4.*
- *Where appropriate, the type-examination certificate referred to in Annex III.*

6. Application to devices in Class IIa.
In line with Article 11(2), this Annex may apply to products in Class IIa, subject to the following:

Annex VI of the MDD 93/42/EEC: EC Declaration of Conformity (Product quality assurance)

1. The manufacturer must ensure application of the quality system approved for the final inspection and testing of the product, as specified in Section 3 and must be subject to the surveillance referred to in Section 4.
In addition, for products placed on the market in sterile condition, and only for those aspects of the manufacturing process designed to secure and maintain sterility, the manufacturer must apply the provisions of Annex V, Sections 3 and 4.
2. The EC declaration of conformity is the part of the procedure whereby the manufacturer who fulfills the obligations imposed by Section 1 ensures and declares that the products concerned conform to the type described in the EC type-examination certificate and meet the provisions of this Directive which apply to them.
The manufacturer affixes the CE marking in accordance with Article 17 and draws up a written declaration of conformity. This declaration must cover one or more medical devices manufactured, clearly identified by means of product name, product code or other unambiguous reference, and are kept by the manufacturer. The CE marking must be accompanied by the identification number of the notified body which performs the tasks referred to in this Annex.
3. Quality system.
3.1. The manufacturer lodges an application for assessment of his quality system with a notified body.
The application must include:

- *The name and address of the manufacturer.*
- *All the relevant information on the product or product category covered by the procedure.*
- *A written declaration specifying that no application has been lodged with any other notified body for the same products.*
- *The documentation on the quality system.*

- *An undertaking by the manufacturer to fulfill the obligations imposed by the quality system approved.*
- *An undertaking by the manufacturer to keep the approved quality system adequate and efficacious.*
- *Where appropriate, the technical documentation on the types approved and a copy of the EC type-examination certificates.*
- *An undertaking by the manufacturer to institute and keep up to date a systematic procedure to review experience gained from devices in the post-production phase, including the provisions referred to in Annex X, and to implement appropriate means to apply any necessary corrective action. This undertaking must include an obligation for the manufacturer to notify the CA of the following incidents immediately on learning of them:*

i. Any malfunction or deterioration in the characteristics and/or performance of a device, as well as any inadequacy in the labeling or the instructions for use which might lead to or might have led to the death of a patient or user or to a serious deterioration in his state of health;

ii. Any technical or medical reason connected with the characteristics or the performance of a device for the reasons referred to in subparagraph (i) leading to a systematic recall of devices of the same type by the manufacturer.

3.2 Under the quality system, each product or a representative sample of each batch is examined and the appropriate tests defined in the relevant standard(s) referred to in Article 5 or equivalent tests are carried out to ensure that the products conform to the type described in the EC type-examination certificate and fulfill the provisions of this Directive which apply to them. All the elements, requirements, and provisions adopted by the manufacturer must be documented in a systematic and orderly manner in the form of written measures, procedures, and instructions. This quality system documentation must permit uniform interpretation of the quality programmes, quality plans, quality manuals, and quality records.

It must include in particular an adequate description of:

- *The quality objectives and the organizational structure, responsibilities and powers of the managerial staff with regard to product quality.*
- *The examinations and tests that will be carried out after manufacture; it must be possible to trace back the calibration of the test equipment adequately.*
- *The methods of monitoring the efficient operation of the quality system.*
- *The quality records, such as reports concerning inspections, tests, calibration and the qualifications of the staff concerned, etc.*
- *Where the final inspection and testing of the products, or elements thereof, are carried out by a third party, the methods of monitoring the efficient operation of the quality system and in particular, the type and extent of control applied to the third party.*

The aforementioned checks do not apply to those aspects of the manufacturing process designed to secure sterility.

3.3. The notified body audits the quality system to determine whether it meets the requirements referred to in section 3.2. It must presume that quality systems which implement the relevant harmonized standards conform to these requirements.

The assessment team must include at least one member with past experience of assessments of the technology concerned. The assessment procedure must include an inspection on the manufacturer's premises and, in duly substantiated cases, on the premises of the manufacturer's suppliers to inspect the manufacturing processes.

The decision must be notified to the manufacturer. It must contain the conclusions of the inspection and a reasoned assessment.

3.4. The manufacturer must inform the notified body which approved the quality system of any plan for substantial changes to the quality system.

The notified body must assess the changes proposed and verify whether after these changes the quality system will still meet the requirements referred to in Section 3.2.

After receiving the above-mentioned information, it must notify the manufacturer of its decision. This decision must contain the conclusions of the inspection and a reasoned assessment.

4. Surveillance.

4.1. The aim of surveillance is to ensure that the manufacturer duly fulfills the obligations imposed by the approved quality system.

4.2. The manufacturer must allow the notified body access for inspection purposes to the inspection, testing, and storage locations and supply it with all relevant information, in particular:

- *The documentation on the quality system.*
- *The technical documentation.*
- *The quality records, such as inspection reports, test data, calibration data, qualification reports of the staff concerned, etc.*

4.3. The notified body must periodically carry out appropriate inspections and assessments to make sure that the manufacturer applies the quality system and must supply the manufacturer with an assessment report.

4.4. In addition, the notified body may pay unannounced visits to the manufacturer. At the time of such visits, the notified body may, where necessary, carry out or ask for tests in order to check that the quality system is working properly and that the production conforms to the requirements of the Directive which apply to it. To this end, an adequate sample of the final products, taken on site by the notified body, must be examined and the appropriate tests defined in the relevant standard(s) referred to in Article 5 or equivalent tests must be carried out. Where one or more of the samples fails to conform, the notified body must take the appropriate measures.

It must provide the manufacturer with an inspection report and, if a test has-been carried out, with a test report.

5. Administrative provisions.

5.1. The manufacturer or his authorized representative must, for a period ending at least five years, and in the case of implantable devices at least 15 years, after the last product has been manufactured, make available to the national authorities:

- *The declaration of conformity.*
- *The documentation referred to in the seventh indent of Section 3.1.*
- *The changes referred to in Section 3.4.*
- *The decisions and reports from the notified body as referred to in the final indent of Section 3.4 and in Sections 4.3 and 4.4.*
- *Where appropriate, the certificate of conformity referred to in Annex III.*

6. Application to devices in Class IIa.

In line with Article 11(2), this Annex may apply to products in Class IIa, subject to the following:

6.1. By way of derogation from Sections 2, 3.1 and 3.2, by virtue of the declaration of conformity the manufacturer ensures and declares that the products in Class IIa are manufactured in conformity with the technical documentation referred to in Section 3 of Annex VII and meet the requirements of this Directive which apply to them.

6.2. For devices in Class IIa the notified body shall assess, as part of the assessment in Section 3.3, the technical documentation as described in Section 3 of Annex VII for at least one representative sample for each device subcategory for compliance with the provisions of this Directive.

6.3. In choosing representative sample(s) the notified body shall take into account the novelty of the technology, similarities in design, technology, manufacturing, and sterilization methods, the intended use and the results of any previous relevant assessments (e.g., with regard to physical, chemical or biological properties) that have been carried out in accordance with this directive. The notified body shall document and keep available to the competent authority its rationale for the sample(s) taken.

6.4. Further samples shall be assessed by the notified body as part of the surveillance assessment referred to in Section 4.3.

Annex VII of the MDD 93/42/EEC: EC Declaration of Conformity

1. The EC declaration of conformity is the procedure whereby the manufacturer or his authorized representative who fulfills the obligations imposed by Section 2 and, in the case of products placed on the market in a sterile condition and devices with a measuring function, the obligations imposed by Section 5 ensures and declares that the products concerned meet the provisions of this Directive which apply to them.

2. The manufacturer must prepare the technical documentation described in Section 3. The manufacturer or his authorized representative must make this documentation, including the declaration of conformity, available to the national authorities for inspection purposes for a period ending at least five years after the last product has been manufactured. In the case of implantable devices, the period shall be at least 15 years after the last product has been manufactured.

3. The technical documentation must allow assessment of the conformity of the product with the requirements of the Directive. It must include in particular:

- *A general description of the product, including any variants planned and its intended use(s).*
- *Design drawings, methods of manufacture envisaged and diagrams of components, sub-assemblies, circuits, etc.*
- *The descriptions and explanations necessary to understand the above-mentioned drawings and diagrams and the operations of the product.*
- *The results of the risk analysis and a list of the standards referred to in Article 5, applied in full or in part, and descriptions of the solutions adopted to meet the ER of the Directive if the standards referred to in Article 5 have not been applied in full.*
- *In the case of products placed on the market in a sterile condition, description of the methods used and the validation report.*
- *The results of the design calculations and of the inspections carried out, etc.; if the device is to be connected to other device(s) in order to operate as intended, proof must be provided that it conforms to the ER when connected to any such device(s) having the characteristics specified by the manufacturer.*
- *The solutions adopted as referred to in Annex I, Chapter I, Section 2.*
- *The pre-clinical evaluation.*
- *The clinical evaluation in accordance with Annex X.*
- *The label and instructions for use.*

4. The manufacturer shall institute and keep up to date a systematic procedure to review experience gained from devices in the post-production phase, including the provisions referred to in Annex X, and to implement appropriate means to apply any necessary corrective actions, taking account of the nature and risks in relation to the product. He shall notify the CA of the following incidents immediately on learning of them:

i. Any malfunction or deterioration in the characteristics and/or performance of a device, as well as any inadequacy in the labeling or the instructions for use which might lead to or might have led to the death of a patient or user or to a serious deterioration in his state of health;

ii. Any technical or medical reason connected with the characteristics on the performance of a device for the reasons referred to in subparagraph (i) leading to systematic recall of devices of the same type by the manufacturer.

5. With products placed on the market in sterile condition and Class I devices with a measuring function, the manufacturer must observe not only the provisions laid down in this Annex but also one of the procedures referred to in Annex II, IV, V or VI. Application of the above-mentioned Annexes and the intervention by the notified body is limited to:

- *In the case of products placed on the market in sterile condition, only the aspects of manufacture concerned with securing and maintaining sterile conditions.*
- *In the case of devices with a measuring function, only the aspects of manufacture concerned with the conformity of the products with the metrological requirements.*

Section 6.1 of this Annex is applicable.

6. Application to devices in Class IIa.

In line with Article 11 (2), this Annex may apply to products in Class IIa, subject to the following derogation:

6.1. Where this Annex is applied in conjunction with the procedure referred to in Annex IV, V or VI, the declaration of conformity referred to in the above-mentioned Annexes forms a single declaration. As regards the declaration based on this Annex, the manufacturer must ensure and declare that the product design meets the provisions of this Directive which apply to it.

Annex VIII of the MDD 93/42/EEC: Statement Concerning Devices for Special Purposes

1. For custom-made devices or for devices intended for clinical investigations the manufacturer or his authorized representative must draw up the statement containing the information stipulated in Section 2.

2. The statement must contain the following information:

2.1. For custom-made devices:

- *The name and address of the manufacturer.*
- *Data allowing identification of the device in question.*
- *A statement that the device is intended for exclusive use by a particular patient, together with the name of the patient.*
- *The name of the medical practitioner or other authorized person who made out the prescription and, where applicable, the name of the clinic concerned.*
- *The specific characteristics of the product as indicated by the prescription.*
- *A statement that the device in question conforms to the ER set out in Annex I and, where applicable, indicating which ER have not been fully met, together with the grounds;*

2.2. For devices intended for the clinical investigations covered by Annex X:

- *Data allowing identification of the device in question.*
- *The clinical investigation plan.*
- *The investigator's brochure.*
- *The confirmation of insurance of subjects.*
- *The documents used to obtain informed consent.*
- *A statement indicating whether or not the device incorporates, as an integral part, a substance or human blood derivative referred to in Section 7.4 of Annex I.*
- *A statement indicating whether or not the device is manufactured utilizing tissues of animal origin as referred to in Directive 2003/32/EC.*
- *The opinion of the ethics committee concerned and details of the aspects covered by its opinion.*
- *The name of the medical practitioner or other authorized person and of the institution responsible for the investigations.*
- *The place, starting date and scheduled duration for the investigations.*
- *A statement that the device in question conforms to the ER apart from the aspects covered by the investigations and that, with regard to these aspects, every precaution has been taken to protect the health and safety of the patient.*

3. The manufacturer must also undertake to keep available for the competent national authorities:

3.1. For custom-made devices, documentation, indicating manufacturing site(s) and allowing an understanding of the design, manufacture, and performances of the product, including the expected performances, so as to allow assessment of conformity with the requirements of this Directive.

The manufacturer must take all the measures necessary to ensure that the manufacturing process produces products which are manufactured in accordance with the documentation mentioned in the first paragraph;

3.2. For devices intended for clinical investigations, the documentation must contain:

- *A general description of the product and its intended use.*
- *Design drawings, methods of manufacture envisaged, in particular as regards sterilization, and diagrams of components, sub-assemblies, circuits, etc.*
- *The descriptions and explanations necessary to understand the above-mentioned drawings and diagrams and the operation of the product.*
- *The results of the risk analysis and a list of the standards referred to in Article 5, applied in full or in part, and descriptions of the solutions adopted to meet the ER of this Directive if the standards referred to in Article 5 have not been applied.*

- *xIf the device incorporates, as an integral part, a substance or human blood derivative referred to in Section 7.4 of Annex I, the data on the tests conducted in this connection which are required to assess the safety, quality, and usefulness of that substance or human blood derivative, taking account of the intended purpose of the device.*
- *If the device is manufactured utilizing tissues of animal origin as referred to in Directive 2003/32/EC, the risk management measures in this connection which have been applied to reduce the risk of infection.*
- *The results of the design calculations and of the inspections and technical tests carried out, etc.*
- *The manufacturer must take all the measures necessary to ensure that the manufacturing process produces products which are manufactured in accordance with the documentation referred to in the first paragraph of this Section.*
- *The manufacturer must authorize the assessment, or audit where necessary, of the effectiveness of these measures.*

4. The information contained in the declarations concerned by this Annex shall be kept for a period of time of at least five years. In the case of implantable devices, the period shall be at least 15 years.

5. For custom-made devices, the manufacturer must undertake to review and document experience gained in the post-production phase, including the provisions referred to in Annex X, and to implement appropriate means to apply any necessary corrective action. This undertaking must include an obligation for the manufacturer to notify the CA of the following incidents immediately on learning of them and the relevant corrective actions:

i. Any malfunction or deterioration in the characteristics and/or performance of a device, as well as any inadequacy in the labeling or the instructions for use which might lead to or might have led to the death of a patient or user or to a serious deterioration in his state of health;

ii. Any technical or medical reason connected with the characteristics or performance of a device for the reasons referred to in subparagraph (i) leading to systematic recall of devices of the same type by the manufacturer.

3.7. HOW TO COMPLY WITH THE MEDICAL DEVICE DIRECTIVE (MDD) 93/42/EEC

In summary, showing compliance with the MDD can be summarized as follows:

Before proceeding with an assessment procedure, it is important to establish whether the manufacturer, can assess the product by themselves or whether there is a need to involve a Notified Body. In order to do that, the manufacturer must first determine under which class the medical devices falls. The involvement of a Notified Body is not necessary for medical devices of Class I unless they have a measuring function or are placed on the market in a sterile condition. For all medical devices belonging to class III, and for medical devices belonging to class IIa and IIb on a representative basis, the design of the medical device and its compliance with the ER must be examined by a Notified Body. The Notified Body issues then, a certificate that indicates, by referring to one of the Annexes II to VI of the MDD, what has been verified. In the case of devices falling within Class I (low risk), other than devices which are custom-made or intended for clinical investigations, the manufacturer performs the conformity assessment and documents the assessment on their own. In this case, the manufacturer shall follow the procedure referred to in Annex VII and draw up the EC declaration of conformity required before placing the device on the market.

In the case of devices falling within Class IIa (medium risk), other than devices which are custom-made or intended for clinical investigations, the manufacturer performs the conformity assessment and documents the assessment in his own right. And, in order to affix the CE marking, shall follow the procedure relating to the EC declaration of conformity set out in Annex VII, coupled with either:

a. The procedure relating to the EC verification set out in Annex IV;

b. The procedure relating to the EC declaration of conformity set out in Annex V (production quality assurance); or

c. The procedure relating to the EC declaration of conformity set out in Annex VI (product quality assurance).

Instead of applying these procedures, the manufacturer may also follow the procedure relating to the EC declaration of conformity set out in Annex II (full quality assurance).

In the case of devices falling within Class IIb (medium risk), other than devices which are custom-made or intended for clinical investigations, the manufacturer shall, in order to affix the CE marking, either:

a. Follow the procedure relating to the EC declaration of conformity set out in Annex II (full quality assurance); in this case, point 4 of Annex II is not applicable; or

b. Follow the procedure relating to the EC type-examination set out in Annex III, coupled with:

i. The procedure relating to the EC verification set out in Annex IV;

ii. The procedure relating to the EC declaration of conformity set out in Annex V (production quality assurance); or

iii. The procedure relating to the EC declaration of conformity set out in Annex VI (product quality assurance).

In the case of devices falling within Class III (high risk), other than devices which are custom-made or intended for clinical investigations, the manufacturer shall, in order to affix the CE marking, either:

a. Follow the procedure relating to the EC declaration of conformity set out in Annex II (full quality assurance); or

b. Follow the procedure relating to the EC type-examination set out in Annex III, coupled with:

i. The procedure relating to the EC verification set out in Annex IV; or

ii. The procedure relating to the EC declaration of conformity set out in Annex V (production quality assurance).

3.8. UNDERSTANDING WHICH PRODUCTS ARE WITHIN THE SCOPE OF THE MEDICAL DEVICE DIRECTIVE (MDD) 93/42/EEC

Article 1 of the MDD defines a 'medical device' to mean any instrument, apparatus, appliance, software, material or other article, whether used alone or in combination, including the software intended by its manufacturer to be used specifically for diagnostic and/or therapeutic purposes and necessary for its proper application, intended by the manufacturer to be used for human beings for the purpose of:

- Diagnosis, prevention, monitoring, treatment or alleviation of disease;
- Diagnosis, monitoring, treatment, alleviation of or compensation for an injury or handicap;
- Investigation, replacement or modification of the anatomy or of a physiological process;
- Control of conception; and which does not achieve its principal intended action in or on the human body by pharmacological, immunological, or metabolic means, but which may be assisted in its function by such means;

In addition, the scope includes an 'accessory,' which means an article which whilst not being a device is intended specifically by its manufacturer to be used together with a device to enable it to be used.

Article 1 of the MDD 93/42/EEC: Definitions, Scope

1. This Directive shall apply to medical devices and their accessories. For the purposes of this Directive, accessories shall be treated as medical devices in their own right. Both medical devices and accessories shall hereinafter be termed devices.

2. For the purposes of this Directive, the following definitions shall apply:

a. 'medical device' means any instrument, apparatus, appliance, software, material or other article, whether used alone or in combination, including the software intended by its manufacturer to be used specifically for diagnostic and/or therapeutic purposes and necessary for its proper application, intended by the manufacturer to be used for human beings for the purpose of:

- *Diagnosis, prevention, monitoring, treatment or alleviation of disease.*
- *Diagnosis, monitoring, treatment, alleviation of or compensation for an injury or handicap.*
- *Investigation, replacement or modification of the anatomy or of a physiological process.*
- *Control of conception, and which does not achieve its principal intended action in or on the human body by pharmacological, immunological or metabolic means, but which may be assisted in its function by such means;*

b. 'Accessory' means an article which whilst not being a device is intended specifically by its manufacturer to be used together with a device to enable it to be used in accordance with the use of the device intended by the manufacturer of the device;

c. 'In vitro diagnostic medical device' means any medical device which is a reagent, reagent product, calibrator, control material, kit, instrument, apparatus, equipment or system, whether used alone or in combination, intended by the manufacturer to be used in vitro for the examination of specimens, including blood and tissue donations, derived from the human body, solely or principally for the purpose of providing information:

· *Concerning a physiological or pathological state;*
· *Concerning a congenital abnormality;*
· *To determine the safety and compatibility with potential recipients; or*
· *To monitor therapeutic measures.*

Specimen receptacles are considered to be in vitro diagnostic medical devices. 'Specimen receptacles' are those devices, whether vacuum-type or not, specifically intended by their manufacturers for the primary containment and preservation of specimens derived from the human body for the purpose of in vitro diagnostic examination.

Products for general laboratory use are not in vitro diagnostic medical devices unless such products, in view of their characteristics, are specifically intended by their manufacturer to be used for in vitro diagnostic examination;

d. 'custom-made device' means any device specifically made in accordance with a duly qualified medical practitioner's written prescription which gives, under his responsibility, specific design characteristics and is intended for the sole use of a particular patient.

The above-mentioned prescription may also be made out by any other person authorized by virtue of his professional qualifications to do so.

Mass-produced devices which need to be adapted to meet the specific requirements of the medical practitioner or any other professional user shall not be considered to be custom-made devices;

e. 'device intended for clinical investigation' means any device intended for use by a duly qualified medical practitioner when conducting investigations as referred to in Section 2.1 of Annex X in an adequate human clinical environment.

For the purpose of conducting clinical investigation, any other person who, by virtue of his professional qualifications, is authorized to carry out such investigation shall be accepted as equivalent to a duly qualified medical practitioner;

f. 'manufacturer' means the natural or legal person with responsibility for the design, manufacture, packaging, and labeling of a device before it is placed on the market under his own name, regardless of whether these operations are carried out by that person himself or on his behalf by a third party.

The obligations of this Directive to be met by manufacturers also apply to the natural or legal person who assembles, packages, processes, fully refurbishes and/or labels one or more ready-made products and/or assigns to them their intended purpose as a device with a view to their being placed on the market under his own name. This subparagraph does not apply to the person who, while not a manufacturer within the meaning of the first subparagraph, assembles or adapts devices already on the market to their intended purpose for an individual patient;

g. 'intended purpose' means the use for which the device is intended according to the data supplied by the manufacturer on the labeling, in the instructions and/or in promotional materials;

h. 'placing on the market' means the first making available in return for payment or free of charge of a device other than a device intended for clinical investigation, with a view to distribution and/or use on the Community market, regardless of whether it is new or fully refurbished;

i. 'putting into service' means the stage at which a device has been made available to the final user as being ready for use on the Community market for the first time for its intended purpose;

j. 'authorized representative' means any natural or legal person established in the Community who, explicitly designated by the manufacturer, acts, and may be addressed by authorities and bodies in the Community instead of the manufacturer with regard to the latter's obligations under this Directive;

k. 'clinical data' means the safety and/or performance information that is generated from the use of a device. Clinical data are sourced from:

- *Clinical investigation(s) of the device concerned;*
- *Clinical investigation(s) or other studies reported in the scientific literature, of a similar device for which equivalence to the device in question can be demonstrated; or*
- *Published and/or unpublished reports on other clinical experience of either the device in question or a similar device for which equivalence to the device in question can be demonstrated;*

l. 'Device subcategory' means a set of devices having common areas of intended use or common technology;
m. 'Generic device group' means a set of devices having the same or similar intended uses or commonality of technology allowing them to be classified in a generic manner not reflecting specific characteristics;
n. 'Single use device' means a device intended to be used once only for a single patient.
3. Where a device is intended to administer a medicinal product within the meaning of Article 1 of Directive 2001/83/EC (1), that device shall be governed by this Directive, without prejudice to the provisions of Directive 2001/83/EC with regard to the medicinal product.
If, however, such a device is placed on the market in such a way that the device and the medicinal product form a single integral product which is intended exclusively for use in the given combination and which is not reusable, that single product shall be governed by Directive 2001/83/EC. The relevant ER of Annex I to this Directive shall apply as far as safety and performance-related device features are concerned.
4. Where a device incorporates, as an integral part, a substance which, if used separately, may be considered to be a medicinal product within the meaning of Article 1 of Directive 2001/83/EC and which is liable to act upon the body with action ancillary to that of the device, that device shall be assessed and authorized in accordance with this Directive.
4a. Where a device incorporates, as an integral part, a substance which, if used separately, may be considered to be a medicinal product constituent or a medicinal product derived from human blood or human plasma within the meaning of Article 1 of Directive 2001/83/EC and which is liable to act upon the human body with action ancillary to that of the device, hereinafter referred to as a 'human blood derivative,' that device shall be assessed and authorized in accordance with this Directive.
5. This Directive shall not apply to:
a. In vitro diagnostic devices;

b. Active implantable devices covered by Directive 90/385/EEC;
c. Medicinal products covered by Directive 2001/83/EC. In deciding whether a product falls under that Directive or this Directive, particular account shall be taken of the principal mode of action of the product;
d. Cosmetic products covered by Directive 76/768/EEC (1);
e) human blood, blood products, plasma or blood cells of human origin or to devices which incorporate at the time of placing on the market such blood products, plasma or cells, with the exception of devices referred to in paragraph 4a;
f. Transplants or tissues or cells of human origin nor to products incorporating or derived from tissues or cells of human origin, with the exception of devices referred to in paragraph 4a;
g. Transplants or tissues or cells of animal origin, unless a device is manufactured utilizing animal tissue which is rendered non-viable or on-viable products derived from animal tissue.
6. Where a device is intended by the manufacturer to be used in accordance with both the provisions on personal protective equipment in Council Directive 89/686/EEC (1) and this Directive, the relevant basic health and safety requirements of Directive 89/686/EEC shall also be fulfilled.
7. This Directive is a specific Directive within the meaning of Article 1(4) of Directive 2004/108/EC of the European Parliament and of the Council (2).
8. This Directive shall not affect the application of Council Directive 96/29/Euratom of 13 May 1996 laying down basic safety standards for the protection of the health of workers and the general public against the dangers arising from ionizing radiation (3), nor of Council Directive 97/43/Euratom of 30 June 1997 on health protection of individuals against the dangers of ionizing radiation in relation to medical exposure.

The medical devices directive outlines the minimum requirements a manufacturer should meet when aiming to ensure the safety and performance characteristics for medical devices in the European market. It covers the following principles:

- The ER: this part of the directive outlines the necessary precautions and requirements to be considered in the design, manufacture, use, and disposal of medical devices.
- Classification of devices and conformity assessment procedures which are based on the level of risk inherent in the device.
- Control of production to ensure conformity to type.
- Monitoring and vigilance.

Manufacturers of Medical Devices should demonstrate conformity to all requirements listed within the Directive in addition to the requirements in subsequent guideline documents such as MEDDEVs (MDDs).

For many of the common techniques and product types there exist what is known as harmonized standards. While use of a harmonized standard is not always mandatory, it is highly recommended as it represents best practice and technical state of the art, which can be used as a presumption of conformity to the relevant part of the directive globally. A manufacturer may however choose to not follow a harmonized standard, which is also okay but, in this case, they must be able to demonstrate clearly how the product meets the relevant safety or performance requirements.

Article 5 of the MDD 93/42/EEC: Reference to standards
1. Member States shall presume compliance with the ER referred to in Article 3 in respect of devices which are in conformity with the relevant national standards adopted pursuant to the harmonized standards the references of which have been publishes in the Official Journal of the European Communities; Member States shall publish the references of such national standards. *2. For the purposes of this Directive, reference to harmonized standards also includes the monographs of the European Pharmacopoeia notably on surgical sutures and on interaction between medicinal products and materials used in devices containing such medicinal products, the references of which have been published in the Official Journal of the European Communities.*

3. If a Member State or the Commission considers that the harmonized standards do not entirely meet the ER referred to in Article 3, the measures to be taken by the Member States with regard to these standards and the publication referred to in paragraph 1 of this Article shall be adopted by the procedure defined in Article 6 (2).

3.9. UNDERSTANDING WHICH PRODUCTS ARE NOT COVERED BY THE MEDICAL DEVICE DIRECTIVE (MDD)

The directive does not apply to the following:

a. *In vitro* diagnostic devices;

b. Active implantable devices covered by Directive 90/385/EEC;

c. Medicinal products covered by Directive 65/65/EEC, including medicinal products derived from blood as covered by Directive 89/381/EEC;

d. Cosmetic products covered by Directive 76/768/EEC;

e. Human blood, blood products, plasma or blood cells of human origin or to devices which incorporate at the time of placing on the market such blood products, plasma or cells, with the exception of devices referred to in paragraph 4a;

f. Transplants or tissues or cells neither of human origin nor to products incorporating or derived from tissues or cells of human origin;

g. Transplants or tissues or cells of animal origin, unless a device is manufactured utilizing animal tissue which is rendered non-viable or non-viable products derived from animal tissue.

3.10. PURPOSE OF THE MEDICAL DEVICE DIRECTIVE (MDD) 93/42/EEC

The Medical Devices Directive was enforced to provide a harmonized regulatory environment for all medical devices sold within the European Union. All products which fall within the scope of the Directive must meet certain essential safety and administrative requirements before being CE marked to show compliance. Such products may then be freely sold within the EU without being subject to any additional national regulations.

3.11. SCOPE OF THE MEDICAL DEVICE DIRECTIVE (MDD) 93/42/EEC

The medical devices directive is one of three directives which together cover all medical equipment. The remaining two directives are: The active implantable medical devices directive (AIMDD) and the *in vitro* diagnostic devices directive (IVDD).

Let's look at how the Directive defines medical devices. According to the MDD: a medical device is defined as:

"any instrument, apparatus, appliance, material or other article, whether used alone or in combination, including the software necessary for its proper application intended by the manufacturer to be used for human beings for the purpose of:

- Diagnosis, prevention, monitoring, treatment or alleviation of disease.
- Diagnosis, monitoring, treatment, or alleviation of or compensation for an injury or handicap.
- Investigation, replacement or modification of the anatomy or of a physiological process.
- Control of conception and which does not achieve its principal intended action in or on the human body by pharmacological, immunological, or metabolic means, but which may be assisted in its function by such means."

The scope of the MDD is quite wide and in some instances, there may be some form of overlap with other Directives. The MDD does however include provisions for all relevant hazards such as mechanical and electrical safety as well as provisions for electromagnetic compatibility. Devices which fall within the scope of the MDD, the AIMDD, and the IVDD are generally excluded from the scope of other Directives. It is also important for a manufacturer to realize that some devices/equipment may fall into or out of the scope of the MDD solely on the basis of the purpose to which the equipment is intended to be used and the way it is marketed. In simpler terms, if a manufacturer makes any claims that a piece of equipment has some form of therapeutic effect or which alleviates the symptoms of any disability, injury or illness, then it falls within the scope of the MDD and a product registration is required as well as appropriate documentation. Common examples of products which are considered to be medical devices include:

- Medicine measuring cups;
- Syringes;
- Dental instruments;
- Stethoscopes;
- Thermometers;
- Prescription spectacles and contact lenses;
- Bandages and splints;
- Dental treatment chairs;
- Wheelchairs;
- Condoms;
- First aid kits.

CHAPTER

4

The Medical Device Regulation (EU MDR 2017/745)

CONTENTS

4.1. WHAT IS THE MEDICAL DEVICE REGULATION (EU MDR 2017/745 OR MDR)?

The medical device regulation (MDR) was published on May 26, 2017, with the aim of replacing the medical device directive (MDD) and the Active Implantable Medical Devices Directive. The MDR will replace the current EU MDD (93/42/EEC) and the EU's Directive on active implantable medical devices (90/385/EEC). All currently certified medical devices and active implantable medical devices must be recertified such that they comply with the new requirements of the MDR. The MDR was meant to come into effect on May 26, 2020, but this has now been delayed for a year to May 2021. The *In-vitro* Diagnostic Regulation (IVDR) replaced the IVD Directive, and comes into effect on May 26, 2022. The new MDR provides an additional time after the date of application allowing placing new products under the MDD for a grace period.

4.2. MAJOR DIFFERENCES BETWEEN THE MEDICAL DEVICE DIRECTIVE (MDD) 93/42/EEC AND MEDICAL DEVICE REGULATION (MDR) 2017/745

The basic structure and requirements of the Regulation remain the same as that of the MDD however a summary of the major differences between the MDD 93/42/EEC and the MDR 2017/745 is provided below:

- There is a wider and more detailed scope of EU legislation which has been extended to include a greater variety of devices such as implants for aesthetic purposes.
- There is much stronger supervision of notified bodies (NBs) by national authorities.
- The rights and responsibilities for manufacturers, authorized representatives (AR), importers, and distributors have been defined more clearly.
- There is an extended database on medical devices (known as Eudamed) which aims to provide a more comprehensive and public information source on devices that are available on the EU market.

- More stringent rules for clinical investigations on devices.
- Introduction of classification rules that divide the broad range of *in vitro* diagnostic medical devices into four different risk classes as it already exists for other medical devices.
- Creation of a medical device coordination group composed of members representing national competent authorities (CA) in the field of medical devices to ensure better coordination between member states.

CHAPTER

5

Medical Device Classification

CONTENTS

5.1. HOW ARE MEDICAL DEVICES REGULATED IN EUROPE?

Placing a device on the EU market means a manufacturer must be able to show it complies with the specific European Medical Device Directives (MDD): AIMDD 90/385/EEC; MDD 93/42/EEC; IVDMDD 98/79/EC. According to the European framework, there are four classes of medical devices: Class I, IIa, IIb, and III (Table 5.1).

Table 5.1. Overview of the Four Classes of Medical Devices as Per the European Framework

Class	Risk Category	Description
I	Low risk	stethoscopes, bandages
Im (measuring device)	Low risk	
Is (sterile device)	Low risk	
IIa	Medium risk	Hearing aids
IIb	Medium to high risk	Ventilators, monitoring equipment
III	High risk	Balloon catheters, prosthetic heart valves

In order to show that a device complies with the essential requirements (ER) of such directives, a CE mark needs to be affixed to it. For this purpose, the device will need to go through the CE marking process, of which the direction depends on the class of the actual device itself and related conformity assessment route. The specific characteristics of the device will determine its actual class and how risky the device is for the patients. The higher the classification, the greater the level of assessment required by notified bodies (NBs). It should be remembered that it is the intended purpose of the device that determines the classification and not the particular technical characteristics. Considerations for classification include the duration of contact with the body, degree of invasiveness and local versus systemic effect.

5.1.1. Class I Medical Devices

Class I medical devices have the lowest perceived risk. Manufacturers of class I medical devices can choose one out of three possible CE marking routes depending on the type of class I device. In general, this is dependent on:

- Is the device sterile? (i.e., a personal protection kit);
- Does it have a measuring function? (i.e., a stethoscope);
- Is the device some other type of class I device? (i.e., not sterile or measuring).

Table 5.2 provides a simplified overview of CE marking routes of class I medical devices:

Table 5.2. Simplified Overview of CE Marking Routes for Class I Medical Devices

Sterile Medical Devices	Measuring Medical Devices	Other Medical Devices
1. Prepare technical documentation to support the declaration of conformity. 2. Notified body assessment product manufacturing aspects related to securing and maintaining sterile conditions. 3. Placing the registration number of the notified body alongside the CE mark. 4. Affixing the CE mark to the product and storing the declaration of conformity and supporting evidence of the competent authority inspection. 5. Placing the medical device on the market	1. Preparing technical documentation to support the declaration of conformity. 2. Notified body conformity assessment of the product with metrological requirements. 3. Compiling a declaration of conformity. 4. Registration with the Competent Authority. 5. Affixing the CE mark to the product and storing the Declaration of Conformity and supporting evidence of the Competent Authority inspection. 6. Placing the medical device on the market	1. Preparing technical documentation to support the declaration of conformity. 2. Compiling a declaration of conformity. 3. Registration with the Competent Authority. 4. Affixing the CE mark to the product, and storing the declaration of conformity and supporting evidence of the competent authority inspection. 5. Placing the medical device on the market

5.1.2. Class IIa Medical Devices

Medical devices of class IIa include surgical gloves, hearing aids, and diagnostic ultrasound machines to name a few. These devices fall into the low to medium risk category. Patients should use them for a short-term period, less than 30 days. Manufacturers of class IIa medical devices will have to back up their declaration of compliance with a Notified body assessment before placing the product on the market. There are four possible routes to CE marking a class IIa product and this is broken down into two groups depending on the product type (i.e., if it is sterile or not). Table 5.3 provides a simplified overview of CE marking routes of class IIa medical devices.

Table 5.3. Simplified Overview of CE Marking Routes for Class IIa Medical Devices

All Medical Devices		**Only Non-Sterile Medical Devices**	
Route 1	**Route 2**	**Route 1**	**Route 2**
1. Full quality assurance audit by a NB. 2. Creation of the declaration of conformity. 3. Affixing the CE mark and NB number. 4. Placing the medical device on the market	1. Preparing technical documentation to support the declaration of conformity. 2. Production quality assurance audit by a NB. 3. Creating a declaration of conformity. 4. Affixing the CE mark and notified body number on the product. 5. Placing the medical device on the market	1. Preparing the technical documentation to support the declaration of conformity. 2. Inspection quality assurance audit by a notified body (not including design and manufacturing). 3. Declaration of Conformity. 4. Affixing the CE mark and notified body number on the product. 5. Placing the medical device on the market	1. Preparing the technical documentation to support the declaration of conformity. 2. A notified body needs to verify every device/batch. 3. Creating a Declaration of Conformity. 4. Affixing the CE mark and notified body number on the product. 5. Placing the medical device on the market

5.1.3. Class IIb Medical Devices

Class IIb medical devices include long-term devices such as contact lenses, surgical lasers and defibrillators. These devices are medical to high-risk and patients can use them for periods of longer than 30 days. Class IIb devices follow similar CE marking routes to class IIa devices but the choice of CE marking route will again depend on the type of product.

Table 5.4 provides a simplified overview of CE marking routes of class IIb medical devices:

Table 5.4. Simplified Overview of CE Marking Routes for Class IIb Medical Devices

All Medical Devices		Only Non-Sterile Medical Devices	
Route 1	**Route 2**	**Route 1**	**Route 2**
1. Full quality assurance audit by a NB. 2. Creation of the declaration of conformity. 3. Affixing the CE mark and NB number on the product. 4. Placing the medical device on the market	1. Type examination by a NB. 2. Production quality assurance audit by a NB. 3. Creating a declaration of conformity. 4. Affixing the CE mark and notified body number on the product. 5. Placing the medical device on the market	1. Type examination by a NB. 2. A notified body inspection quality assurance audit (not including design and manufacturing). 3. Declaration of Conformity. 4. Affixing the CE mark and notified body number on the product. 5. Placing the medical device on the market	1. Type examination by a NB. 2. A notified body needs to verify every device/batch. 3. Creating a Declaration of Conformity. 4. Affixing the CE mark and notified body number on the product. 5. Placing the medical device on the market

5.1.4. Class III Medical Devices

Class III medical devices have the highest risk profile and require permanent monitoring throughout their lifecycle. Common examples include cardiovascular catheters, hip-joint implants, and heart valves to name a few. A typical CE marking route may include an audit of the technical documentation and a quality system/product inspection as well as focus on one or more aspects of the device design and production.

Table 5.5 provides a simplified overview of CE marking routes of class III medical devices:

Table 5.5. Simplified Overview of CE Marking Routes for Class III Medical Devices

All Medical Devices			**Only Non-Sterile Medical Devices**
Route 1	**Route 2**	**Route 3**	**Route 1**
1. Full quality assurance audit by a notified body. 2. Creating a declaration of conformity. 3. Affixing the CE mark and notified body number on the product. 4. Placing the medical device on the market	1. Design dossier examination by a notified body. 2. Creating a declaration of conformity. 3. Affixing the CE mark and notified body number on the product. 4. Placing the medical device on the market	1. Type examination by a notified body. 2. Production quality assurance audit by a notified body (not including design). 3. Declaration of conformity. 4. Affixing the CE mark and notified body number of the product. 5. Placing the medical device on the market	1. Type examination by a notified body. 2. A notified body needs to verify every device/batch. 3. Creating a declaration of conformity. 4. Affixing the CE mark and notified body number on the product. 5. Placing the medical device on the market

5.2. EU GUIDELINES ON MEDICAL DEVICE CLASSIFICATION

Detailed explanations of the European classification systems can be found in MEDDEV 2.4/1. General requirements of this classification scheme are such that all devices must:

- Meet the ER irrespective of the class of the device.
- Be subject to the reporting requirements under the medical device vigilance system.
- Be CE marked (except for custom-made devices and devices that are intended for clinical investigation).

These medical device classification systems are needed to channel medical devices into the proper conformity assessment route (Table 5.6) as subjecting all existing medical devices to rigorous conformity assessment procedures would just not be feasible economically.

Table 5.6. Conformity Assessment Procedures vs. Classes

Conformity Assessment Procedures	**Classes**					
Annexes	**I**	**I Sterile**	**I Measuring**	**IIA**	**IIB**	**III**
II (+ sect. 4)						√
II (– sect 4)				√	√	
III					√	√
IV		√	√	√	√	√
V		√	√	√	√	√
VI		√	√	√	√	
VII	√	√	√	√		

When carrying out a classification as per MEDDEV 2.4/1, the manufacturer should first decide if the product is a medical device as defined by Directive 93/42 or if it is an accessory to a medical device. Determining the exact grouping of the device will aid a manufacturer in knowing which directive to apply.

Active implantable devices and devices for *in vitro* diagnosis are covered by separate directives which in actual fact do not apply the classification rules from guidelines supplied by MEDDEV 2.4/1 thereby further indicating that different groups of devices will utilize different rules.

Classification rules as defined by MEDDEV 2.4/1 are based on terms related to: (1) the duration of contact with the patient, (2) degree of invasiveness, and (3) the part of the body affected by the use of the device itself.

One of the first steps to applying the classification rules is understanding the basic definitions used in the guideline, all of which are defined in Section I of Annex IX of the Directive.

5.2.1. Duration

The duration of contact with the patient, also known as *time* can be classified into transient, short term or long term. Transient is defined as that which is normally intended for continuous use for *less than 60 minutes* while short term is defined as that which is normally intended for continuous use for *not more than 30 days* and long term is that which is normally intended for continuous use *for more than 30 days*.

5.2.2. Continuous Use

In addition, understanding the concept of continuous use is important when the intended use of the device is taken into consideration. The above-mentioned concept of duration (transient, short term and long term) are all defined in terms of continuous use which is better defined as an uninterrupted actual use for the intended purpose of the device in question. For example, a scalpel which may be used on a patient for cutting of tissue will not normally last for more than a few seconds at a time and can therefore be defined as a continuous transient use device. On the other hand, a device that is used and replaced immediately either by the same or identical device is considered as discontinued and shall be defined as an extension of the continuous use of the device. A common example would be a suture needle which is used for stitching wounds.

5.2.3. Invasiveness

Invasive devices are defined as a device which either, completely or partly, penetrates inside the body. This usually occurs either through a body orifice (better known as any natural opening in the body) or through the surface of the body. An invasive device which penetrates inside the body through the surface of the body, with the aid or in the context of a surgical operation is better known as a surgically invasive device. Thus, a surgically invasive device always implies that it enters through an artificially created opening, which could be a large opening, such as a surgical incision, or as small as a needle hole created by a suture needle.

5.2.4. Implantable Device

An implantable device is defined as any device which is intended for complete introduction into the human body or which will be replaced by an epithelial surface or the surface of the eye; either by surgical intervention and which is intended to remain in place after the procedure. In addition, any device intended to be partially introduced into the human body through surgical intervention and intended to remain in place after the procedure for at least 30 days is also considered an implantable device. When defining an implantable device, the idea behind the term 'procedure' must be taken into consideration and as such, an implantable device must remain in the patient after the procedure is complete. As such, a procedure must include the surgical activity during which the implantable device is placed in the body and the immediate post-operative procedures associate with the initial procedure itself. That being said, the term procedure does not include the removal of an implant, which in effect must be considered to be another procedure. As a result, the placing of an implant and the removal of the implant would be become two different surgical procedures. Furthermore, some partially implanted devices are also classified to be implants but clear distinction is needed.

5.2.5. Active Devices

Active medical devices are defined as any medical device in which their operation depends on a source of electrical energy or any power source other than that directly generated by the human body or gravity and which acts by converting this energy. Those medical devices that aim to simply transmit energy, substances, or other elements between an active medical device and the patient themselves, without any significant change, are not considered to be active medical devices. In order for this definition to be conveyed properly, the concept of "act by converting energy" and "significant changes" must be clearly explained and understood. To "act by converting energy" includes the conversion of energy in the device and/or conversion at the interface between the device itself and the tissues or within the tissues. The concept of "significant changes" on the other hand includes changes in the nature, level, and density of energy. For example, an electrode would not be defined as an active medical device as per this definition as long as the energy input is intended to be the same as the energy output. On the other hand, electrodes used in electro surgery for cutting tissues can be defined as active medical devices as per the above definition as their operation depends on energy provided by a generator and their action is achieved by conversion of the energy at the interface between the device and the tissue itself or within the tissue. However, those electrodes that are intended for E.C.G or E.E.G's are to normally define as active devices because they do not normally act by conversion of energy. Manufacturers should however consult the "Guidelines relating to the application of the Council Directive 90/385/EEC on active implantable medical devices (MEDDEV 2.1/2)" should they require more information or better understanding on the issue.

5.2.6. Devices with a Measuring Function

Devices are considered to have a measuring function if they measure a parameter such as volume or a physiological parameter and display or indicate its value in a defined unit or measurement. Common examples include urine bags or thermometers.

Certain criteria have been laid out as per MEDDEV 2.1/5 (1998) which a Guidance document relating to the application of medical devices with a measuring function. This guidance document states criteria which if fulfilled together, indicate that a device does, in fact, have a measuring function. These criteria are defined below:

1. The device is intended by the manufacturer to measure:
   - Quantitatively a physiological or anatomical parameter.
   - A quantity or a quantifiable characteristic of energy or of substances delivered to or removed from the human body.
2. The result of the overall measurement:
   - Is displayed in legal units or other acceptable units within the meaning of Directive 80/181/ECC.
   - Is compared to at least one point of reference indicated in legal units or other acceptable units in compliance with the pre-mentioned directive.
3. The intended purpose implies accuracy (claimed either explicitly or implicitly) where any form of non-compliance with the implied accuracy could result in a significant adverse effect on the patient's health and safety.

Note 1: the phrase "claimed implicitly" covers cases where the user, on the basis of the designation of the device or of its accompanying documents, or on the basis of the common use is entitled to expect accuracy where the accuracy of the measurement has an impact on the diagnosis or therapy of the patient (Table 5.7).

Table 5.7. Common Examples of Devices with a Measuring Function

Criteria	**Example**
Devices used for measuring body temperature.	
Pacifiers which include a temperature display including those with only a change of color where criteria b is met.	
• Devices for indicating that body temperature is above or below a specified value.	
• Non-active non-invasive devices for measuring blood pressure	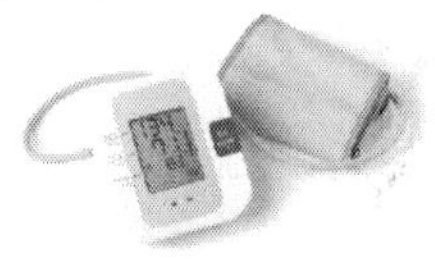
• Non-active devices for measuring intra-ocular pressure	
• Devices for measuring volume or pressure or flow of liquid or gases delivered to or removed from the human body. This includes any container with a graduation scale or with a single point graduation where criteria c is met.	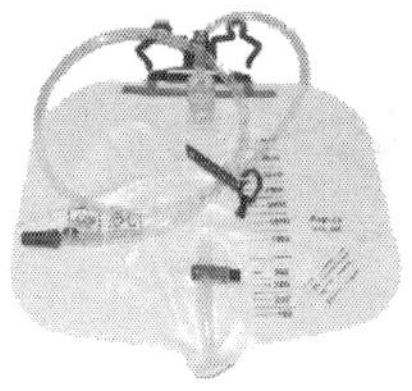

5.2.7. Devices without a Measuring Function

Common examples of devices without a measuring function:

Criteria	Example

• Devices used for the delivery of liquid to the human body (for example medicine spoons, cups, droppers, without graduation or scale or display of measuring unit).	
• Devices used for displaying trends of physiological parameters (for example urine bags without graduation or scale, caliper for obesity).	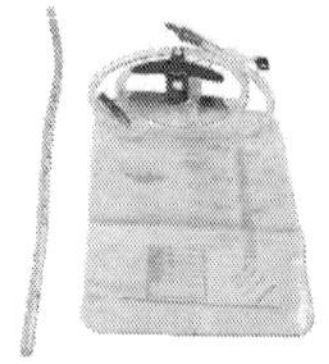
• Eye-test charts	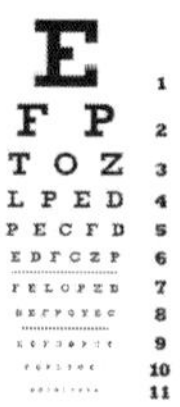

5.3. UNDERSTANDING HOW TO APPLY THE RULES OF MEDICAL DEVICE CLASSIFICATION

When applying the classification rules, the following should be considered:

- The intended purpose of the device as opposed to technical characteristics as this will determine the class of the device itself (unless these have a direct influence on the intended purpose).
- The intended use must not be confused with common accidental uses of the device.
- The intended purpose assigned by the actual manufacturer to the specific device should determine the class of the device rather than allowing for classification by inferences such as the class of other similar products determining the class of a specific device.
- Classification of device components in relation to the system as a whole or individually should be taken into consideration. For example, combination devices with parts that have different functional purposes may be analyzed separately with respect to each of the parts; such as a drainage device, that has an invasive tube and a non-invasive collection device. Each of these

components may be classified separately.

- Accessories must be classified separately from their parent devices.
- If a device can be classified by several rules, then the highest possible class should apply.
- If a device is not intended to be used solely in a specific part of the body, then it must be considered and classified based on the most critical specified use. As such, classification of the device will have to be determined based on claims contained in the information supplied with the medical device itself. For this reason, manufacturers must be extremely specific with regards to the overall intended use while also providing as a minimum requirement, either appropriate positive or negative indications for use.
- Multi-application equipment (such as laser printers or identification cameras) which may be used in combination with medical devices is not actual medical devices unless their manufacturer places them on the market with specific intended purpose as a medical device or for use with medical devices only.
- Software that is regarded as driving or influencing the use of a medical device may fall in the same class of the device which it influences. If, however, such standalone software is not regarded as driving or influencing the use of a medical device then it should be classified separately and in its own right.

5.4. HOW TO APPLY THE EU MEDICAL DEVICE CLASSIFICATION RULES

In order to appropriately apply the EU medical device rules for classification, manufacturers must take into account all the rules laid out in the guidelines to ensure their devices are properly classified. In addition, all device characteristics or combination of characteristics in relation to its intended purpose must be taken into consideration. One such classification example is defined below:

A wound drainage device produced by Manufacturer X needs to be classified:

A simple wound drainage device has three components that must be taken into consideration, namely the cannula, the tubing, and the collector unit. If the particular device in question is to be sold without a cannula, then the classification of the cannula does not need to be taken into consideration.

The manufacturer has also stated that the device will be used under short term duration, i.e., uninterrupted intended use will be for more than 60 minutes and less than 30 days. It also assumed that any collected liquids are not intended to be re-infused into the body nor reprocessed for eventual re-infusion and that the device is not intended to be connected to a powered suction system (Table 5.8).

Table 5.8. Intended Uses of Devices and Their Risk Classification

Intended Uses	Rule	Class
Surgically invasive cannula to reach a wound site in the pleural cavity to drain the cavity	7	IIA
Non-invasive tubing to evacuate body liquids towards the collector	1	I
Non-invasive collector to receive the body liquids	1	I

From the above-defined specifications by the manufacturer, it is clear that they would have a choice of applying a Class II A risk level to the whole device. Alternatively, they would have to carry out a separate conformity assessment procedure for the cannula and another conformity assessment procedure for the tubing and another for the collector.

As defined by MEDDEV 2.4/1, different rules will apply to different groups of devices (Table 5.9). As a rule of thumb, however, it should be remembered that the characteristics or combination of characteristics in accordance with its intended purpose of the device that determines the highest class will determine the overall class for the device (Figures 5.1–5.5).

Table 5.9. Different Rules That Apply for Risk Classification of Different Devices

Type of device	Applicable Rules as per MEDDEV 2.4/1 Classification Rules
Non-invasive	1,2,3,4
Invasive	5,6,7,8
Active	9,10,11,12
Special	13,14,15,16,17,18

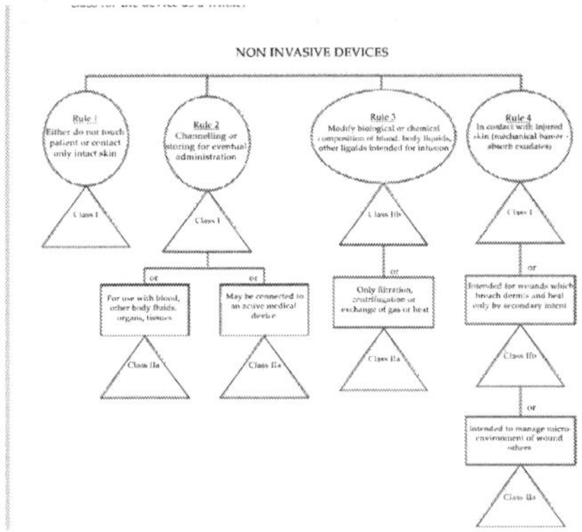

Figure 5.1. Flow diagram showing risk classification routes based on rules 1–4 from MEDDEV2.4/1.

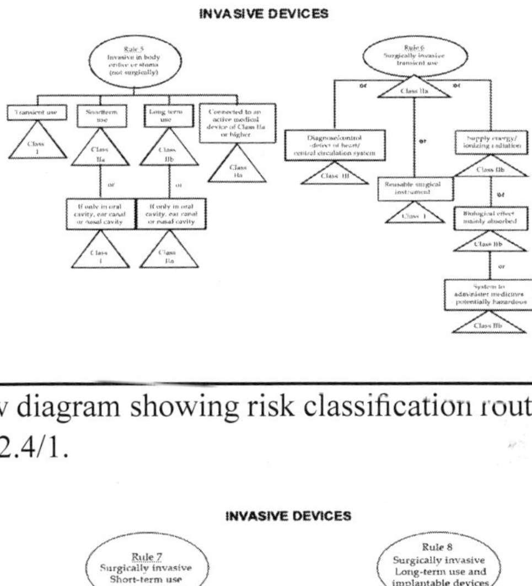

Figure 5.2. Flow diagram showing risk classification routes based on rules 5–6 from MEDDEV2.4/1.

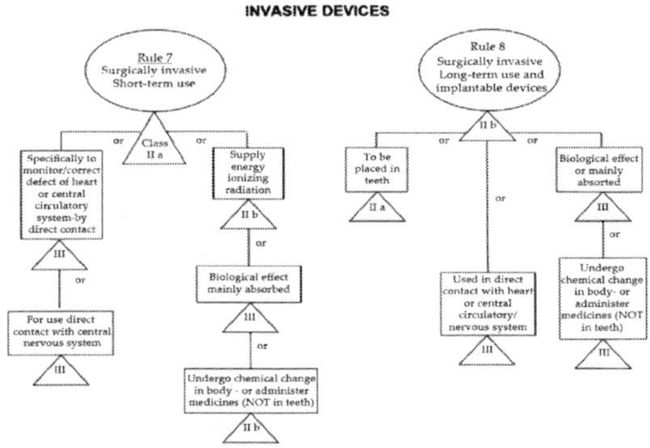

Figure 5.3. Flow diagram showing risk classification routes based on rules 7–8 from MEDDEV2.4/1.

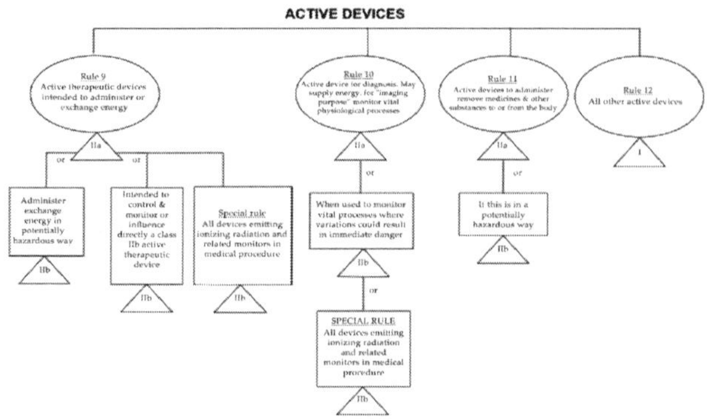

Figure 5.4. Flow diagram showing risk classification routes based on rules 9–12 from MEDDEV2.4/1.

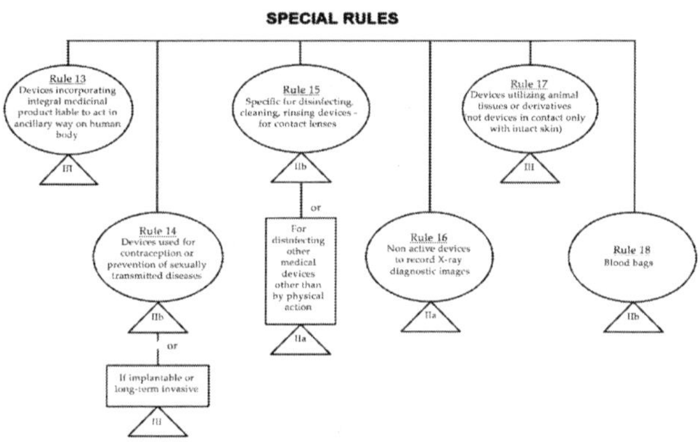

Figure 5.5. Flow diagram showing risk classification routes based on rules 13–17 from MEDDEV2.4/1.

5.5. UNDERSTANDING THE RULES GOVERNING GUIDELINES FOR THE CLASSIFICATION OF MEDICAL DEVICES ACCORDING TO MEDDEV 2.4/1

The below sub-chapter covers a general explanation of the rules and practical issues governing the EU guidelines for medical device classification as laid out in MEDDEV 2.4/1.

5.5.1. Rule 1: Devices That Either Do Not Touch the Patient or Contact Intact Skin Only

General explanation of the rule: Rule 1 forms a fallback rule that is applicable to all medical devices that are note generally covered by a more specific rule later. It applies in general to all devices that meet only intact skin, and which does not touch the patient (Table 5.10).

Table 5.10. Classification Rules as Per Rule 1 of MEDDEV 2.4/1

Rule 1	Examples
All non-invasive devices are in Class I, unless one of the rules set out hereinafter applies.	• Body liquid collection devices intended to be used in such a way that a return flow is unlikely (e.g., to collect body wastes such as urine collection bottles, ostomy pouches, incontinence pads or collectors used with wound drainage devices). They may be connected to the patient by means of catheters and tubing. • Devices used to immobilize body parts and/ or to apply force or compression on them (e.g., non-sterile dressings used to aid the healing of a sprain, plaster of Paris, cervical collars, gravity traction devices, compression hosiery). • Devices intended in general for external patient support (e.g., hospital beds, patient hoists, walking aids, wheelchairs, stretchers, dental patient chairs). • Corrective glasses, frames, stethoscopes for diagnosis, eye occlusion plasters, incision drapes, conductive gels, non-invasive electrodes (electrodes for EEG or ECG), and image intensifying screens. • Permanent magnets for removal of ocular debris

5.5.1.1. Understanding the Practical Issues around Rule 1 Classification

Some non-invasive devices are indirectly in contact with the human body and can influence internal physiological processes by storing, channeling, or treating blood, other body liquids or liquids which are returned or infused into the body or by generating energy that is delivered to the body. These must be excluded when applying this rule and should be handled by another rule because of the hazards inherent in such direct influence on the body.

5.5.2. Rule 2: Channeling or Storing for Eventual Administration

General explanation of the rule: these types of devices must be considered separately from the non-contact devices of rule 1 because they may be indirectly invasive. They channel or store substances that will be eventually delivered into the body. Typically, these devices after used in transfusion, infusion, extracorporeal circulation, delivery of anesthetic gases and oxygen. In some cases, devices covered under this rule are very simple devices that are gravity activated (Table 5.11).

Table 5.11. Classification Rules as Per Rule 2 of MEDDEV 2.4/1

Rule 2	**Examples**
• All non-invasive devices intended for channeling or storing blood, body liquids or tissues, liquids or gases for the purpose of eventual infusion, administration or introduction into the body are in Class IIa: • If they may be connected to an active medical device in Class IIa or a higher class	• Devices intended to be used as channels in active drug delivery systems, e.g., tubing intended for use with an infusion pump. • Devices used for channeling, e.g., antistatic tubing for anesthesia, anesthesia breathing circuits and pressure indicator, pressure limiting devices. • Syringes for infusion pumps.
• If they are intended for use for storing or channeling blood or other body liquids or for storing organs, parts of organs or body tissues (are in Class II a)	• Devices intended to channel blood (e.g., in transfusion, extracorporeal circulation). • Devices intended for temporary storage and transport of organs for transplantation. • Devices intended for long term storage of biological substances and tissues such as corneas, sperm, human embryos, etc.

• In all other cases they are in Class I	• Devices that provide a simple channeling function, with gravity providing the force to transport the liquid, e.g., administration sets for infusion. • Devices intended to be used for a temporary containment or storage function such as cups and spoons specifically intended for administering medicines. • Syringes without needle.

5.5.2.1. Understanding the Practical Issues around Rule 2 Classification

If a device can be used for a purpose that would result it to be connected to an active device, then such a device will automatically fall in Class II A unless the manufacturer clearly states that it should not be connected to an active device of Class II A or higher. It should also be noted that within Rule 2, blood bags are covered as an exception under a separate rule (Rule 18).

Special concepts are also noted in the above table for Rule 2 and are explained as follows:

Note 1: All non-invasive devices intended for channeling or storing blood, body liquids or tissues, liquids or gases for the purpose of eventual infusion, administration or introduction into the body are in Class IIa: *If they may be connected to an active medical device in Class IIa or a higher class.*

Such connection is deemed to exist between a non-active device and an active device where a non-active device forms a link in the transfer of the substance between the patient and the active device. Furthermore, the safety and performance of one of the devices is influenced by the other device.

An example as explained in the guideline would be tubing in an extracorporeal circulation system which is downstream from a blood pump and in the same blood flow circuit but not directly in contact with the pump.

Note 2: "Devices intended to be used for a temporary containment or storage function such as cups and spoons specifically intended for administering medicines."

This statement refers to solutions intended for preservation of organs during storage and transport and not medical devices.

5.5.3. Rule 3: Devices That Modify Biological or Chemical Composition of Blood, Body Liquids or Other Liquids

General explanation of the rule: these types of devices must be considered separately from non-contact devices of rule 1 because they are indirectly invasive and treat or modify substances that will eventually be delivered into the body. Rule 3 covers mostly the more sophisticated elements of extracorporeal circulation sets, dialysis systems and auto transfusion systems as well as devices for extracorporeal treatment of body fluids which may not be reintroduced immediately into the body, including where the patient is not in a closed loop with the device (Table 5.12).

Table 5.12. Classification Rules as Per Rule 3 of MEDDEV 2.4/1

Rule 3	Examples
• All non-invasive devices intended for modifying the biological or chemical composition of blood, other body liquids or other liquids intended for infusion into the body are in Class IIb	• Devices intended to remove undesirable substances out of the blood by exchange of solutes such as hem dialyzers. • Devices intended to separate cells by physical means, e.g., gradient medium for sperm separation. • Haemodialysis concentrates
• Unless the treatment consists of filtration, centrifugation or exchange of gas or heat, in which case they are in Class IIa	
• Unless the treatment consists of filtration, centrifugation or exchange of gas or heat, in which case they are in Class IIa	• Particulate filtration of blood in an extracorporeal circulation system. These are used to remove particles and emboli from the blood. • Centrifugation of blood to prepare it for transfusion or auto transfusion. • Removal of carbon dioxide from the blood and/or adding oxygen. • Warming or cooling the blood in an extracorporeal circulation system

5.5.3.1. Understanding the Practical Issues of Rule 3 Classification

These devices are normally used in conjunction with an active medical device covered under rule 9 or rule 11. Filtration and centrifugation should be understood in the context of this rule as exclusively mechanical methods.

5.5.4. Rule 4: Devices in Contact With Injured Skin

General explanation of the rule: Rule 4 is intended to cover mainly wound dressings regardless/independent of the wound. Traditional types of products such as those commonly used as mechanical barriers, are often well understood and do not result in any great hazard. In addition, there have been rapid technological developments in this area, with the emergence of new types of wound dressings for which non-traditional claims are made. Examples of non-traditional claims include management of the micro-environment of a wound to enhance its natural healing mechanisms. Other ambitious claims may also be included, and these involve the mechanism of healing by secondary intent. This rule also includes some devices used on breached dermis which may even have a life-sustaining or lifesaving purpose, such as when there is full thickness destruction of the skin over a large area and/or systemic effect. Furthermore, dressings containing medicinal products acting as ancillary to the dressing fall within Class III under Rule 13 (Table 5.13).

Table 5.13. Classification Rules as Per Rule 4 of MEDDEV 2.4/1

Rule 4	Examples
• All non-invasive devices which come into contact with injured skin: o Are in Class I if they are intended to be used as a mechanical barrier, for compression or for absorption of exudates,	• Wound dressings, such as absorbent pads, island dressings, cotton wool, wound strips and gauze dressings to act as a barrier or to maintain the wound positionally or to absorb exudates from the wound.
o Are in Class IIb if they are intended to be used principally with wounds which have breached the dermis and can only heal by secondary intent	• Are principally intended to be used with severe wounds that have substantially and extensively breached the dermis, and where the healing process can only be by secondary intent such as: o Dressings for chronic extensive ulcerated wounds. o Dressings for severe burns having breached the dermis and covering an extensive area. o Dressings for severe wounds. o Dressings incorporating means of augmenting tissue and providing a temporary skin substitute

o Are in Class IIa in all other cases, including devices principally intended to manage the micro-environment of a wound.	• Have specific properties intended to assist the healing process by controlling the level of moisture at the wound during the healing process and to generally regulate the environment in terms of humidity and temperature, levels of oxygen and other gases and pH values or by influencing the process by other physical means. • These devices may specify additional healing properties whilst not being intended for extensive wounds requiring healing by secondary intent. • Adhesives for topical use. • Polymer film dressings, hydrogel dressings and non-medicated impregnated gauze dressings.

5.5.4.1. Understanding the Practical Issues of Rule 4 Classification

The products that fall within this rule are extremely claim sensitive, for example, a polymeric film dressing would fall in Class II A if the intended use is to manage the micro-environment of the wound and in Class I if its intended use is limited to retaining an invasive cannula at the wound site. Thus, the actual use of the device would determine the class and it is impossible to say that a dressing is in a given class without knowing its intended use as defined by the manufacturer. So, for example, a claim that the device is interactive or active with respect to the wound healing process usually implies that the device is in Class II B.

It should also be noted that most dressings intended for a function/use that is in Class II A or II B will also perform functions that are in Class I and as a result, such devices are nevertheless classed according to the intended use in the higher class.

For such devices incorporating medicines rule 13 should be consulted and for devices incorporating animal tissues rule 17 should be consulted.

Special concepts defined by this rule include:

1. **Breached Dermis:** Whereby the wound exposes at least partly the subcutaneous tissues.
2. **Secondary Intent:** Whereby the wound heals by first being filled with granulation tissue. This is followed by the epithelium which

grows back over the granulation tissue and the wound contracts. This is in contrast to primary intent which implies that the edges of the wound are close enough or pulled together, e.g., by suturing, to allow the wound to heal.

5.5.5. Rule 5: Devices Invasive in Body Orifices

General explanation of the rule: Rule 5 is intended to cover invasiveness with respect to the body orifices such as the ear, mouth, nose, eye, anus, urethra, and vagina. This must be considered separately from invasiveness that penetrates through a cut in the body surfaces, for example through surgical invasiveness. A further distinction must be made between invasiveness with respect to the less vulnerable anterior parts of the ear, mouth, and nose and the other anatomical sites that can be accessed through natural body orifices. The surgically created stoma, which for example allows the evacuation of urine or faces, should also be considered as a body orifice. Devices covered by this rule tend to be diagnostic and therapeutic instruments used in particular specialties (ENT, ophthalmology, dentistry, proctology, urology, and gynecology) (Table 5.14).

Table 5.14. Classification Rules as Per Rule 5 of MEDDEV 2.4/1

Rule 5	**Examples**
• All invasive devices with respect to body orifices, other than surgically invasive devices and which are not intended for connection to an active medical device: • Are in Class I if they are intended for transient use? • Are in Class IIa if they are intended for short term use. • Except if they are used in the oral cavity as far as the pharynx, in an ear canal up to the eardrum or in a nasal cavity, in which case they are in Class I. • Are in Class IIb if they are intended for long term use.	• Handheld mirrors used in dentistry to aid in dental diagnosis and surgery, dental impression materials, tubes used for pumping the stomach, impression tray, enema devices, examination gloves and prostatic balloon dilation catheters. • Contact lenses, urinary catheters, tracheal tubes, stents, vaginal pessaries and perineal re-education devices. • Dressings for nose bleeds, dentures intended to be removed by the patient. • Urethral stents.

• Except if they are used in the oral cavity as far as the pharynx, in an ear canal up to the ear drum or in a nasal cavity and are not liable to be absorbed by the mucous membrane, in which case they are in Class IIa. • All invasive devices with respect to body orifices, other than surgically invasive devices, intended for connection to an active medical device in Class IIa or a higher class, are in Class IIa.	• Orthodontic wire, fixed dental prostheses, fissures sealants. • Tracheostomy or tracheal tubes connected to a ventilator, blood oxygen analyzers placed under the eyelid, powered nasal irrigators, nasopharyngeal airways, some enteral feeding tubes, fiber optics in endoscopes connected to surgical lasers, suction catheters or tubes for stomach drainage, dental aspirator tips.

Rule 5	**Examples**
• All invasive devices with respect to body orifices, other than surgically invasive devices and which are not intended for connection to an active medical device: • Are in Class I if they are intended for transient use? • Are in Class IIa if they are intended for short term use. • Except if they are used in the oral cavity as far as the pharynx, in an ear canal up to the eardrum or in a nasal cavity, in which case they are in Class I. • Are in Class IIb if they are intended for long term use. • Except if they are used in the oral cavity as far as the pharynx, in an ear canal up to the ear drum or in a nasal cavity and are not liable to be absorbed by the mucous membrane, in which case they are in Class IIa. • All invasive devices with respect to body orifices, other than surgically invasive devices, intended for connection to an active medical device in Class IIa or a higher class, are in Class IIa.	• Handheld mirrors used in dentistry to aid in dental diagnosis and surgery, dental impression materials, tubes used for pumping the stomach, impression tray, enema devices, examination gloves and prostatic balloon dilation catheters. • Contact lenses, urinary catheters, tracheal tubes, stents, vaginal pessaries and perineal re-education devices. • Dressings for nose bleeds, dentures intended to be removed by the patient. • Urethral stents. • Orthodontic wire, fixed dental prostheses, fissures sealants. • Tracheostomy or tracheal tubes connected to a ventilator, blood oxygen analyzers placed under the eyelid, powered nasal irrigators, nasopharyngeal airways, some enteral feeding tubes, fiber optics in endoscopes connected to surgical lasers, suction catheters or tubes for stomach drainage, dental aspirator tips.

5.5.6. Rule 6: Surgically Invasive Devices for Transient Use

General explanation of the rule: This rule covers principally three major groups of devices, namely 1) devices that are used to create a conduit through the skin such as needles; 2) surgical instruments such as scalpels and saws; 3) various types of catheters and/or suckers and so on (Table 5.15).

Table 5.15. Classification Rules as Per Rule 6 of MEDDEV 2.4/1

Rule 6	Examples
• All surgically invasive devices intended for transient use are in Class IIa unless they are: ○ Intended specifically to diagnose, monitor or correct a defect of the heart or of the central circulatory system through direct contact with these parts of the body, in which case they are in Class III. ○ Reusable surgical instruments, in which case they are in Class I. ○ Intended to supply energy in the form of ionizing radiation in which case they are in Class IIb. ○ Intended to have a biological effect or to be wholly or mainly absorbed in which case they are in Class IIb. ○ Intended to administer medicines by means of a delivery system, if this is done in a manner that is potentially hazardous considering of the mode of application, in which case they are Class IIb.	• Needles used for suturing, needles or syringes, lancets, suckers, single use scalpels, single use scalpel blades, support devices in ophthalmic surgery, staplers, surgical swabs, drill bits connected to active devices, surgical gloves, etchants, tester of artificial heart valves, heart valve sizers and holders, trial hip prosthesis heads or stems, swabs to sample exudates, single use aortic punches. • Cardiovascular catheters (e.g., angioplasty balloon catheters), including related guide wires and dedicated disposable cardiovascular surgical instruments, e.g., electrophysiological catheters, electrodes for electrophysiological diagnosis and ablation. • Catheters containing or incorporating sealed radioisotopes, where the radioactive isotope as such is not intended to be released into the body, if used in the central circulatory system. • Scalpels, scalpel handles, drill bits, saws, that are not intended for connection to an active device, and retractors forceps, excavators, and chisels.

	• Catheters containing or incorporating sealed radioisotopes, where the radioactive isotope as such is not intended to be released into the body, if used in the circulatory system, excluding the central circulatory system. • Devices for repeated self-application where dosage levels and the nature of the medicinal product are critical, e.g., insulin pens.

5.5.6.1. Explanation of Special Concepts Applicable in Rule 6

- ***Note 1:*** Terms such as *"surgically invasive device," "central circulatory system"* and *"reusable surgical instruments"* are defined in Section I of Annex IX of the Directive. Surgical instruments that are connected to an active device are not considered to be "reusable surgical instruments."
- ***Note 2:*** The phrase "correct a defect" does not cover devices that are used as an accessory in heart surgery, for example clamps. The first indent of this rule does not apply to aortic punches and similar cutting instruments which perform a similar function to that of a scalpel.
- ***Note 3:*** This applies to surgical instruments which are not specifically intended for purposes described in the first indent of the table, and irrespective of the site of application, are 1) in class IIA if they are intended for single use and 2) in class I if they are reusable.
- ***Note 4:*** *Dedicated* means that the intended purpose of the device is to diagnose, monitor or correct a defect of the heart or of the central circulatory system.
- ***Note 5:*** The phrase *"biological effect"* pertains to all materials and devices that have the potential to affect tissues following use in a surgically invasive procedure. A material is considered to have some kind of biological effect if it actively and intentionally induces, alters, or prevents a response from the tissues that is mediated by specific reactions at a molecular level. These kinds of devices may also be described as bioactive.
- *Wholly or mainly absorbed:* The term absorption refers to the degradation of a material within the body and the metabolic elimination of the resulting degradation products from the body.

- *Note 6:* The concept of *"potentially hazardous manner"* is related to the characteristics of the device and not the competence of the user.

5.5.7. Rule 7: Surgically Invasive Devices for Short-Term Use

General explanation of the rule: Devices covered by rule 7 are most those devices used in the context of surgery or post-operative care such as clamps or drains; infusion devise such as cannulae or needles and catheters of various types (Table 5.16).

Table 5.16. Classification Rules as Per Rule 7 of MEDDEV 2.4/1

Rule 7	Examples
• All surgically invasive devices intended for short term use are in Class IIa unless they are intended: ○ Either specifically to diagnose, monitor or correct a defect of the heart or of the central circulatory system through direct contact with these parts of the body, in which case they are in Class III. ○ Or specifically for use in direct contact with the central nervous system, in which case they are in Class III. ○ Or to supply energy in the form of ionizing radiation in which case they are in Class IIb.	• Clamps, infusion cannulae, skin closure devices, temporary filling materials. • Tissue stabilizers used in cardiac surgery. • Cardiovascular catheters, cardiac output probes and temporary pacemaker leads. • Thoracic catheters intended to drain the heart, including the pericardium. • Carotid artery shunts.
○ Intended to have a biological effect or to be wholly or mainly absorbed in which case they are in Class III. ○ Or to undergo chemical change in the body, except if the devices are placed in the teeth, or to administer medicines, in which case they are Class IIb.	• Neurological catheters, cortical electrodes. • Brachytherapy devices. • Absorbable sutures and biological adhesives. • Adhesives

5.5.7.1. Practical Issues of Rule 7 Classification

Note 1: *"administer medicines"*-the administration of medicines is more than just channeling as it implies also storage and/or influencing the volume and rate of medicine being delivered. It should also be noted that implanted capsules for the slow release of medicines are actually medicines and not medical devices.

Note 2: *"correct a defect"*-this phrase does not cover devices that are used as an accessory in heart surgery, for example tissue stabilizers.

5.5.8. Rule 8: Surgically Invasive Devices for Long-Term Use and Implantable Devices

General explanation of the rule: devices falling under rule 8 are mostly implants in the orthopedic, dental, ophthalmic, and cardiovascular fields as well as soft tissue implants such as implants used in plastic surgery (Table 5.17).

Table 5.17. Classification Rules as Per Rule 8 of MEDDEV 2.4/1

Rule 8	Examples
• All implantable devices and long-term surgically invasive devices are in Class IIb unless they are intended: o To be placed in the teeth in which case they are in Class IIa. o To be used in direct contact with the heart, the central circulatory system or the central nervous system, in which case they are Class III. o To have a biological effect or to be wholly or mainly absorbed, in which case they are in Class III. o Or to undergo chemical change in the body, except if the devices are placed in the teeth, or to administer medicines, in which case they are in Class III.	• Prosthetic joint replacements, ligaments, shunts, stents, nails, plates, intra-ocular lenses, internal closure devices, tissue augmentation implants, infusion ports, peripheral vascular grafts, penile implants, non-absorbable sutures, bone cements and maxillo-facial implants, visco-elastic surgical devices intended specifically for ophthalmic anterior segment surgery. • Bridges, crowns, dental filling materials and pins, dental alloys, ceramics, and polymers. • Prosthetic heart valves, aneurysm clips, vascular prostheses, spinal stents, vascular stents, CNS electrodes and cardiovascular sutures. • Permanent vena cava filters. • Absorbable sutures, adhesives, and implantable devices claimed to be bioactive through the attachment of surface coatings such as phosphorylcholine. • Rechargeable non-active drug delivery systems.

Practical issues of classification for Rule 8

- ***Note 1:*** These products are implants because in normal conditions a significant amount of the substance remains at the surgical site after the procedure. If, however these devices contain animal tissues or derivatives of animal tissues, they are covered by rule 17.
- ***Note 2:*** *"to be placed in the teeth"*
- Implants without bioactive coatings intended to secure teeth or prostheses to the maxillary or mandibular bones are in Class II B following the general rule.
- ***Note 3:*** *"undergo chemical change"*

The clause about chemical change under this rule does not apply to products such as bone cements where the chemical change takes place during the placement and does not continue in long term.

5.5.9. Rule 9: Active Therapeutic Devices Intended to Administer or Exchange Energy

General explanation of the rule: Medical devices classified by Rule 8 are mostly electrical equipment used in surgery such as lasers and surgical generators.

Table 5.18. Classification Rules as Per Rule 9 of MEDDEV 2.4/1

Rule 9	Examples
• All active therapeutic devices intended to administer, or exchange energy are in Class IIa. o Unless their characteristics are such that they may administer or exchange energy to and from the human body in a potentially hazardous way, taking account of the nature, the density and the site of application of the energy, in which case they are in Class IIb.	• Electrical and/or magnetic and electromagnetic energy: Muscle stimulators and external bone growth stimulators, TENS devices and eye electromagnets, electrical acupuncture. • Thermal energy: Cryosurgery equipment, heat exchangers, except the types described below. • Mechanical energy: Powered dermatomes, powered drills, and dental hand pieces. • Light: Phototherapy for skin treatment and for neonatal care. • Sound: Hearing aids. • Ultrasound: Equipment for physiotherapy. • Kinetic energy: Lung ventilators.

	• Thermal energy: Incubators for babies, warming blankets, blood warmers, electrically powered heat exchangers, for instance those used with patient's incapable of reacting, communicating, and/or who are without a sense of feeling. • Electrical energy: High-frequency electrosurgical generators, and electrocautery equipment, including their electrodes, external pacemakers, external defibrillators, electroconvulsive therapy equipment. • Coherent light: Surgical lasers. • Ultrasound: surgical ultrasound devices. • Ionizing radiation: Radioactive sources for after loading therapy, therapeutic cyclotrons, linear accelerators, therapeutic X-ray sources.
• All active devices intended to control and monitor the performance of active therapeutic devices in Class IIb or intended to influence directly the performance of such devices are in Class IIb.	• External feedback systems for active therapeutic devices, after loading control devices

There are, however, devices for specialized treatment such as radiation treatment. In addition, another category consists of stimulation devices, although not all of them can be considered as delivering dangerous levels of energy considering the tissue involved (Table 5.18).

5.5.9.1. Explanation of Special Concepts Pertaining to Rule 9

Note 1: *"Exchange energy to and from the human body in a potentially hazardous way"*

The decision as to whether a medical device administers or exchanges energy to and from the human body in a potentially hazardous way should take various factors into consideration. The concept of "potentially hazardous" is dependent on the type of technology involved and the intended application of the device to the patient, and not on the measures adopted by the manufacturer in view of good design management. That being said, the manufacturer's obligation to comply with design requirements and solutions adopted (such as use of standards) exist independently from the classification system.

5.5.10. Rule 10: Active Devices for Diagnosis

General explanation of the rule: this rule covers primarily a whole range of widely used equipment in the field's ultrasound diagnosis and the capture of physiological signals as well as therapeutic and diagnostic radiology (Table 5.19).

Table 5.19. Classification Rules as Per Rule 10 of MEDDEV 2.4/1

Rule 10	Examples
• Active devices intended for diagnosis are in Class IIa: ○ If they are intended to supply energy which will be absorbed by the human body, except for devices used to illuminate the patient's body, in the visible spectrum. ○ If they are intended to image *in vivo* distribution of radiopharmaceuticals. ○ If they are intended to allow direct diagnosis or monitoring of vital physiological processes. ○ Unless they are specifically intended for monitoring of vital physiological parameters, where the nature of variations is such that it could result in immediate danger to the patient, for instance variations in cardiac performance, respiration, activity of CNS in which case they are in Class IIb. • Active devices intended to emit ionizing radiation and intended for diagnostic and therapeutic interventional radiology including devices which control or monitor such devices, or which directly influence their performance, are in Class II B.	• Magnetic resonance equipment, pulp testers, evoked response stimulators, diagnostic ultrasound. • Gamma cameras, positron emission tomography and single photon emission computer tomography. • Electrocardiographs, electroencephalographs, cardioscopes with or without pacing pulse indicators • Electronic thermometers. • Electronic stethoscopes. • Electronic blood pressure measuring equipment. • Intensive care monitoring and alarm devices (for, e.g., blood pressure, temperature, oxygen saturation), biological sensors, blood gas analyzers used in open heart surgery, cardioscopes, and apnea monitors, including apnea monitors in home care. • Diagnostic X-ray sources.

5.5.10.1. Explanation of Special Concepts Pertaining to Rule 10

Note 1: "vital physiological processes"

These vital physiological processes and parameters include for example;

respiration, heart rate, cerebral functions, blood gases, blood pressure and body temperature. Medical devices intended to be used for continuous surveillance of vital physiological processes in anesthesia, intensive care or emergency care are in Class IIB, whilst medical devices intended to be used to obtain readings of vital physiological signals in routine check-ups and in self-monitoring are in Class IIA. A thermal imaging device intended to monitor blood flow is not considered to be a temperature measuring device.

Note 2: "with or without pacing pulse indicators"

This is applicable for devices specifically intended to monitor AIMDs and which fall under the AIMD Directive.

Note 3: Therapeutic interventional radiology.

This refers to diagnosis being carried out during surgical procedures.

Note 4: "including devices which control or monitor"

This refers to active devices for the control, monitoring or influencing and not to the subsequent processing, recording or viewing of the resulting image.

5.5.11. Rule 11: Active Devices to Administer, Remove Medicines and Other Substances to or From the Body

General explanation of the rule: this rule is intended to cover primarily drug delivery systems and anesthesia equipment (Table 5.20).

Table 5.20. Classification Rules as Per Rule 11 of MEDDEV 2.4/1

Rule 11	Examples
• All active devices intended to administer and/or remove medicines, body liquids or other substances to or from the body are in Class IIa. • ner: that is potentially hazardous, taking account of the nature of the substances involved, of the part of the body concerned and of the mode of application, in which case they are in Class IIb. • Unless this is done in a man	• Infusion pumps, ventilators, anesthesia machines, anesthetic vaporizers, dialysis equipment, blood pumps for heart-lung machines, hyperbaric chambers, pressure regulators for medical gases, medical gas mixers, moisture exchangers in breathing circuits if used on unconscious or non-spontaneously breathing patients.

	• Nebulizers where the failure to deliver the appropriate dosage characteristics could be hazardous.

5.5.12. Rule 12: All Other Active Devices

General explanation of the rule: this forms a fallback rule to cover all active devices not covered by previous rules (Table 5.21).

Table 5.21. Classification Rules as Per Rule 12 of MEDDEV 2.4/1

Rule 12	Examples
• All other active devices are in Class I	• Active diagnostic devices intended to illuminate the patient's body in the visible spectrum such as examination lights or to optically view the body such as surgical microscopes. • Devices intended in general for external patient support (e.g., hospital beds, patient hoists, walking aids, wheelchairs, stretchers, dental patient chairs). • Active diagnostic devices intended for thermography. • Active devices intended for recording, processing or viewing of diagnostic images. • Dental curing lights.

5.6. SPECIAL RULES

5.6.1. Rule 13: Devices Incorporating a Medicinal Substance

General explanation of the rule: This rule is intended to cover combination devices that contain a medicinal substance incorporated into the device. This substance is usually present for the purpose of assisting the functioning of that device. This rule, however, does not cover those devices incorporating substances which under other circumstances may be considered as medicinal substances. It does however cover that which is incorporated into the device exclusively for the purpose of maintaining certain characteristics of the device and which are not liable to act on the body. A common example includes agents for the preservation of solutions for contact lenses. The primary function of the device does not rely on the pharmacological effect of the medicine as it the latter is true, then the product would be classified as being a medicine rather than a devicc. In that case, it will not be covered by the MEDDEV 2.4/1 Directive (Table 5.22).

Table 5.22. Classification Rules as Per Rule 13 of MEDDEV 2.4/1

Rule 13	Examples
• All devices incorporating, as an integral part, a substance which, if used separately, can be considered to be a medicinal product as defined in Article 1 of the Directive 65/65/EEC, and which is liable to act on the human body with action ancillary to that of the devices, are in Class III.	• Antibiotic bone cements, condoms with spermicide, heparin coated catheters, endodontic materials with antibiotics. • Ophthalmic irrigation solutions principally intended for irrigation, which contain components which support the metabolism of the endothelial cells of the cornea-Dressings incorporating an antimicrobial agent where the purpose of such an agent is to provide ancillary action on the wound

***Note 1:** Integral Part.*

Integral part means that the device and the medicinal substance form one physical unit.

5.6.2. Rule 14: Devices Used for Contraception or Prevention of Sexually Transmitted Diseases

General explanation of the rule: the intended uses defined by this rule relates to special cases of human vulnerability that cannot be covered by the normal criteria of time, invasiveness, and organic function. While the rule does cover two very different device applications, some devices may perform both functions. A common example of this would be condoms. It should also be noted that devices intended to prevent the sexual transmission of the HIV are also covered by this rule (Table 5.23).

Table 5.23. Classification Rules as Per Rule 14 of MEDDEV 2.4/1

Rule 14	Examples
• All devices used for contraception or the prevention of the transmission of sexually transmitted diseases are in Class IIb. ○ Unless they are implantable or long-term invasive devices, in which case they are in Class III.	• Condoms, contraceptive diaphragms. • Contraceptive intrauterine devices (IUDs)

Note 1: Contraceptive intrauterine devices (IUDs).

Those intrauterine contraceptives whose primary purpose is to release progestogens are not medical devices.

5.6.3. Rule 15: Specific Disinfecting, Cleaning, and Rinsing Devices

General explanation of the rule: this rule is mainly intended to cover various contact lens fluids as well as those substances used principally in a medical environment to disinfect medical devices (Table 5.24).

Table 5.24. Classification Rules as Per Rule 15 of MEDDEV 2.4/1

Rule 15	Examples
• All devices intended specifically to be used for disinfecting, cleaning, rinsing or, when appropriate hydrating contact lenses are in Class IIb. • All devices intended specifically to be used for disinfecting medical devices are in Class IIa. • This rule does not apply to products that are intended to clean medical devices other than contact lenses by means of physical action	• Contact lens solutions, comfort solutions. • Disinfectants specifically intended for instance for endoscopes or hemodialysis equipment, sterilizers specifically intended to sterilize medical devices in a medical environment and washer disinfectors. • Cleaners which disinfect prosthetic dentures.

5.6.3.1. Practical Issues of Rule 15 Classification

Note 1: This rule does not apply to mechanical means of cleaning of devices, such as brushes and ultrasound. Such products will only fall under this directive if they are specifically intended for use with medical devices.

5.6.4. Rule 16: Non-Active Devices to Record X-Ray Diagnostic Images (Table 5.25)

Table 5.25. Classification Rules as Per Rule 16 of MEDDEV 2.4/1

Rule 16	Examples
Non-active devices specifically intended for recording of X-ray diagnostic images are in Class II a.	X-ray films, photostimulable phosphor plates

Note: This rule refers to primary recording media such as X-ray films and not to media used for subsequent reproduction.

5.6.5. Rule 17: Devices Utilizing Animal Tissue or Derivatives

General explanation of the rule: Rule 17 covers those devices that contain animal tissues, or which are made of animal tissues that have been rendered non-viable. It also considers those derivatives from such tissues where there is no longer any capacity for cellular metabolic activity (i.e., non-viable tissues). It should be noted that those medical devices containing non-inactivated animal tissues and/or any human tissues or derivatives are excluded from the scope of this Directive. It should be further noted that the manufacture of some medical devices may use industrial raw materials, and this may result in the presence of small amounts of tallow or tallow derivatives which are not considered as derivatives of animal tissues for the purpose of this rule. In this case, this rule does not apply (Table 5.26).

Table 5.26. Classification Rules as Per Rule 17 of MEDDEV 2.4/1

Rule 17	Examples
All devices manufactured utilizing animal tissues or derivatives rendered non-viable are Class III except where such devices are intended to come into contact with intact skin only	Biological heart valves, porcine xenograft dressings, catgut sutures, implants, and dressings made from collagen.

Practical classification issues: those medical devices made of non-viable animal tissue that come into contact with intact skin only are in Class I in accordance to rule 1.

Note 1: Animal tissues or derivatives.

Derivatives are defined as those products that are processed from animal tissues and exclude substances such as milk, silk, beeswax, hair, lanolin.

Note 2: Intact skin.

Intact skin includes the skin around an established stoma unless however, this skin is breached.

5.6.6. Rule 18: Blood Bags

General explanation of the rule: This is a special rule that covers only blood bags (Table 5.27).

Table 5.27. Classification Rules as Per Rule 18 of MEDDEV 2.4/1

Rule 18	Examples
By derogation from other rules, blood bags are in Class IIb	Blood bags (including those containing or coated with an anticoagulant). Where blood bags have a function greater than for storing purposes and include systems for preservation other than anti-coagulants then other rules (e.g., rule 13) may apply

5.7. UNDERSTANDING CLASSIFICATION RULES UNDER THE NEW MDR 2017/745

One of the major changes under the EU MDR that medical device manufacturers will see as they transition over to the new regulation is the change in requirements for device classification. It is also important to remember that as this medical device classifications change, so do the requirements imposed on manufacturers and their related interaction with their Notified Body.

Both the MDD and now the EU MDR 2017/745 classify medical devices into four classes (I, IIa, IIb, and III) based on their intended purposes and inherent risks. The higher risk the medical device, the more rules, and regulations it must comply with.

5.8. UNDERSTANDING HOW CATEGORIES OF MEDICAL DEVICES ARE DEFINED UNDER MDR

It is important to note that the duration of use and definition of the different product categories are very similar. There has however, been some changes in vocabulary to further refine the understanding of such terminology. This similarity between the MDD and MDR is important as if there was such a definition change, it would most likely mean a manufacturer has to reconsider their products risk classification and take into consideration the related repercussions on the business.

Chapter I: understanding the definitions specific to classification rules as defined in the MDR 2017/745

Chapter I of the MDR 2017/745: Definitions specific to classification rules

Duration of Use:

- *'Transient' means normally intended for continuous use for less than 60 minutes.*
- *'Short term' means normally intended for continuous use for between 60 minutes and 30 days. 1.*
- *'Long term' means normally intended for continuous use for more than 30 days.*

Invasive and Active Devices:

· *'Body orifice' means any natural opening in the body, as well as the external surface of the eyeball, or any permanent artificial opening, such as a stoma.*

· *'Surgically invasive device' means:*

a. An invasive device which penetrates inside the body through the surface of the body, including through mucous membranes of body orifices with the aid or in the context of a surgical operation; and

b. A device which produces penetration other than through a body orifice.

'Reusable surgical instrument' means an instrument intended for surgical use in cutting, drilling, sawing, scratching, scraping, clamping, retracting, clipping or similar procedures, without a connection to an active device and which is intended by the manufacturer to be reused after appropriate procedures such as cleaning, disinfection, and sterilization have been carried out.

'Active therapeutic device' means any active device used, whether alone or in combination with other devices, to support, modify, replace or restore biological functions or structures with a view to treatment or alleviation of an illness, injury or disability.

'Active device intended for diagnosis and monitoring' means any active device used, whether alone or in combination with other devices, to supply information for detecting, diagnosing, monitoring or treating physiological conditions, states of health, illnesses or congenital deformities.

'Central circulatory system' means the following blood vessels: arteriae pulmonales, aorta ascendens, arcus aortae, aorta descendens to the bifurcatio aortae, arteriae coronariae, arteria carotiscommunis, arteria carotis externa, arteria carotisinterna, arteriae cerebrales, truncus brachiocephalicus, venae cordis, venae pulmonales, vena cava superior and vena cava inferior.

> *'Central nervous system' means the brain, meninges, and spinal cord. 'Injured skin or mucous membrane' means an area of skin or a mucous membrane presenting a pathological change or change following disease or a wound.*

Table 5.28. Comparison of MDR and MDD Definitions

Term	MDR	MDD	Explanation of Changes
Surgically Invasive Device	An invasive device which penetrates inside the body through the surface of the body, including through mucous membranes of body orifices with the aid or in the context of a surgical operation; (Annex VIII)	An invasive device which penetrates inside the body through the surface of the body, with the aid or in the context of a surgical operation.	There is now clarification that surgically invasive also includes penetration through mucous membranes of body orifices
Injured skin or mucous membrane	Means an area of skin or a mucous membrane presenting a pathological change or change following disease or a wound.	n/a	introduced definition for injured skin or mucous membrane
Active device	Covered under Article 2. 'active device' means any device, the operation of which depends on a source of energy other than that generated by the human body for that purpose, or by gravity, and which acts by changing the density of or converting that energy. Devices intended to transmit energy, substances, or other elements between an active device and the patient, without any significant change, shall not be deemed to be active devices. Software shall also be deemed to be an active device;	Any medical device operation of which depends on a source of electrical energy or any source of power other than that directly generated by the human body or gravity and which acts by converting this energy. Medical devices intended to transmit energy, substances, or other elements between an active medical device and the patient, without any significant change, is not considered to be active medical devices. Stand-alone software is considered to be an active medical device.	Device which acts by changing the density of energy are also considered active devices. This is considered a clarification only

Reusable surgical instrument	Instrument intended for surgical use in cutting, drilling, sawing, scratching, scraping, clamping, retracting, clipping or similar procedures, without a connection to an active device and which is intended by the manufacturer to be reused after appropriate procedures such as cleaning, disinfection, and sterilization have been carried out.	Instrument intended for surgical use by cutting, drilling, sawing, scratching, scraping, clamping, retracting, clipping or similar procedures, without connection to any active medical device and which can be reused after appropriate procedures have been carried out.	devices which can be reused vs. intended by the manufacturer to be reused; mostly a clarification
Continuous use	(a) The entire duration of use of the same device without regard to temporary interruption of use during a procedure or temporary removal for purposes such as cleaning or disinfection of the device. Whether the interruption of use or the removal is temporary shall be established in relation to the duration of the use prior to and after the period when the use is interrupted or the device removed; and (b) the accumulated use of a device that is intended by the manufacturer to be replaced immediately with another of the same type.	Means 'an uninterrupted actual use of the device for the intended purpose.' However where usage of a device is discontinued in order for the device to be replaced immediately by the same or an identical device this shall be considered an extension of the continuous use of the device	concept of continuous use extended to include devices that may be temporarily removed to be cleaned or disinfected and then reused; However this is still considered a clarification since such use would be treated as 'continuous use' under the Directive as well
Direct diagnosis	A device is considered to allow direct diagnosis when it provides the diagnosis of the disease or condition in question by itself or when it provides decisive information for the diagnosis.	n/a	new definition

Implantable device	Covered under Article 2.5 'implantable device' means any device, including those that are partially or wholly absorbed, which is intended: to be totally introduced into the human body, or—to replace an epithelial surface or the surface of the eye, by clinical intervention and which is intended to remain in place after the procedure. Any device intended to be partially introduced into the human body by clinical intervention and intended to remain in place after the procedure for at least 30 days shall also be deemed to be an implantable device;	Any device which is intended:—to be totally introduced into the human body or,—to replace an epithelial surface or the surface of the eye, by surgical intervention which is intended to remain in place after the procedure. Any device intended to be partially introduced into the human body through surgical intervention and intended to remain in place after the procedure for at least 30 days is also considered an implantable device.	Devices that are partially or wholly absorbed are considered implantable; 'Clinical' intervention vs. 'surgical' intervention

What might also help a manufacturer is a simple comparison of definitions between the MDR and MDD. Viewing this side by side will make it easier for a manufacturer to decipher the changes between the two regulations (Table 5.28).

Chapter II: Implementing classification rules are defined in Annex VIII of the MDR 2017/745 as follows:

Chapter II of the MDR 2017/745: Implementing Rules
Application of the classification rules shall be governed by the intended purpose of the devices. *If the device in question is intended to be used in combination with another device, the classification rules shall apply separately to each of the devices.* *Accessories for a medical device and for a product listed in Annex XVI shall be classified in their own right separately from the device with which they are used.* *Software, which drives a device or influences the use of a device, shall fall within the same class as the device. If the software is independent of any other device, it shall be classified in its own right.*

If the device is not intended to be used solely or principally in a specific part of the body, it shall be considered and classified on the basis of the most critical specified use. *If several rules, or if, within the same rule, several sub-rules, apply to the same device based on the device's intended purpose, the strictest rule and sub-rule resulting in the higher classification shall apply.* *In calculating the duration referred to in Section 1, continuous use shall mean:* *a. The entire duration of use of the same device without regard to temporary interruption of use during a procedure or temporary removal for purposes such as cleaning or disinfection of the device. Whether the interruption of use or the removal is temporary shall be established in relation to the duration of the use prior to and after the period when the use is interrupted or the device removed; and* *b. The accumulated use of a device that is intended by the manufacturer to be replaced immediately with another of the same type.* *A device is considered to allow direct diagnosis when it provides the diagnosis of the disease or condition in question by itself or when it provides decisive information for the diagnosis.*

While the MDD allows for 18 classification rules, the new MDR contains 22 rules. These 18 rules can be found in Annex IX of the MDD. The MDR on the other hand contains 22 rules, of which the 18 classification rules of the MDD are still present. These 22 rules can be found in Annex VIII of the MDR.

There are however, four more rules than the previous medical device directive (MDD) which is partly due to the broader scope of the MDR. This means the MDR will apply to certain products that are not regulated through the MDD. All of the rules are based on the potential risks associated with the device itself, its technical design as well as how the device itself is manufactured.

The 22 device classification rules are defined in Annex VIII of the MDR 2017/745 as follows:

Chapter III of The MDR 2017/745: Classification Rules
Non-Invasive Devices
Rule 1: *All non-invasive devices are classified as class I, unless one of the rules set out hereinafter applies.*

Rule 2: *All non-invasive devices intended for channeling or storing blood, body liquids, cells or tissues, liquids or gases for the purpose of eventual infusion, administration or introduction into the body are classified as class IIa:* · *If they may be connected to a class IIa, class IIb or class III active device; or* · *If they are intended for use for channeling or storing blood or other body liquids or for storing organs, parts of organs or body cells and tissues, except for blood bags; blood bags are classified as class IIb.* *In all other cases, such devices are classified as class I.*
Rule 3: *All non-invasive devices intended for modifying the biological or chemical composition of human tissues or cells, blood, other body liquids or other liquids intended for implantation or administration into the body are classified as class IIb, unless the treatment for which the device is used consists of filtration, centrifugation or exchanges of gas, heat, in which case they are classified as class IIa.* *All non-invasive devices consisting of a substance or a mixture of substances intended to be used in vitro in direct contact with human cells, tissues or organs taken from the human body or used in vitro with human embryos before their implantation or administration into the body are classified as class III.*
Rule 4: *All non-invasive devices which come into contact with injured skin or mucous membrane are classified as:* · *Class I if they are intended to be used as a mechanical barrier, for compression or for absorption of exudates;* · *Class IIb if they are intended to be used principally for injuries to skin which have breached the dermis or mucous membrane and can only heal by secondary intent;* · *Class IIa if they are principally intended to manage the micro-environment of injured skin or mucous membrane; and* · *Class IIa in all other cases. This rule applies also to the invasive devices that come into contact with injured mucous membrane.*
Invasive Devices

Rule 5: *All invasive devices with respect to body orifices, other than surgically invasive devices, which are not intended for connection to an active device or which are intended for connection to a class I active device are classified as:* · *Class I if they are intended for transient use;* · *Class IIa if they are intended for short-term use, except if they are used in the oral cavity as far as the pharynx, in an ear canal up to the ear drum or in the nasal cavity, in which case they are classified as class I; and* · *Class IIb if they are intended for long-term use, except if they are used in the oral cavity as far as the pharynx, in an ear canal up to the ear drum or in the nasal cavity and are not liable to be absorbed by the mucous membrane, in which case they are classified as class IIa.* *All invasive devices with respect to body orifices, other than surgically invasive devices, intended for connection to a class IIa, class IIb or class III active device, are classified as class IIa.*
Rule 6: *All surgically invasive devices intended for transient use are classified as class IIa unless they:* · *Are intended specifically to control, diagnose, monitor or correct a defect of the heart or of the central circulatory system through direct contact with those parts of the body, in which case they are classified as class III;* · *Are reusable surgical instruments, in which case they are classified as class I;* · *Are intended specifically for use in direct contact with the heart or central circulatory system or the central nervous system, in which case they are classified as class III;* · *Are intended to supply energy in the form of ionizing radiation in which case they are classified as class IIb;* · *Have a biological effect or are wholly or mainly absorbed in which case they are classified as class IIb; or* · *Are intended to administer medicinal products by means of a delivery system, if such administration of a medicinal product is done in a manner that is potentially hazardous taking account of the mode of application, in which case they are classified as class* IIb.

Rule 7: *All surgically invasive devices intended for short-term use are classified as class IIa unless they:* • *Are intended specifically to control, diagnose, monitor or correct a defect of the heart or of the central circulatory system through direct contact with those parts of the body, in which case they are classified as class III;* • *Are intended specifically for use in direct contact with the heart or central circulatory system or the central nervous system, in which case they are classified as class III;* • *Are intended to supply energy in the form of ionizing radiation in which case they are classified as class IIb;* • *Have a biological effect or are wholly or mainly absorbed in which case they are classified as class III;* • *Are intended to undergo chemical change in the body in which case they are classified as class IIb, except if the devices are placed in the teeth; or* • *Are intended to administer medicines, in which case they are classified as class* IIb.
Rule 8: *All implantable devices and long-term surgically invasive devices are classified as class IIb unless they:* • *Are intended to be placed in the teeth, in which case they are classified as class IIa;* • *Are intended to be used in direct contact with the heart, the central circulatory system or the central nervous system, in which case they are classified as class III;* • *Have a biological effect or are wholly or mainly absorbed, in which case they are classified as class III;* • *Are intended to undergo chemical change in the body in which case they are classified as class III, except if the devices are placed in the teeth;* • *Are intended to administer medicinal products, in which case they are classified as class III* • *Are active implantable devices or their accessories, in which cases they are classified as class III;* • *Are breast implants or surgical meshes, in which cases they are classified as class III;*

• *Are total or partial joint replacements, in which case they are classified as class III, with the exception of ancillary components such as screws, wedges, plates, and instruments; or* • · *Are spinal disc replacement implants or are implantable devices that come into contact with the spinal column, in which case they are classified as class III with the exception of components such as screws, wedges, plates, and instruments.*
Active Devices
Rule 9: *All active therapeutic devices intended to administer or exchange energy are classified as class IIa unless their characteristics are such that they may administer energy to or exchange energy with the human body in a potentially hazardous way, taking account of the nature, the density and site of application of the energy, in which case they are classified as class IIb.* *All active devices intended to control or monitor the performance of active therapeutic class IIb devices, or intended directly to influence the performance of such devices are classified as class IIb.* *All active devices intended to emit ionizing radiation for therapeutic purposes, including devices which control or monitor such devices, or which directly influence their performance, are classified as class IIb.* *All active devices that are intended for controlling, monitoring, or directly influencing the performance of active implantable devices are classified as class III.*
Rule 10: *Active devices intended for diagnosis and monitoring are classified as class IIa:* • *If they are intended to supply energy which will be absorbed by the human body, except for devices intended to illuminate the patient's body, in the visible spectrum, in which case they are classified as class I;* • *If they are intended to image in vivo distribution of radiopharmaceuticals; or* • *If they are intended to allow direct diagnosis or monitoring of vital physiological processes, unless they are specifically intended for monitoring of vital physiological parameters and the nature of variations of those parameters is such that it could result in immediate danger to the patient, for instance variations in cardiac performance, respiration, activity of the central nervous system, or they are intended for diagnosis in clinical situations where the patient is in immediate danger, in which cases they are classified as class IIb.*

Active devices intended to emit ionizing radiation and intended for diagnostic or therapeutic radiology, including interventional radiology devices and devices which control or monitor such devices, or which directly influence their performance, are classified as class IIb.

Rule 11: *Software intended to provide information which is used to take decisions with diagnosis or therapeutic purposes is classified as class IIa, except if such decisions have an impact that may cause:*

- *Death or an irreversible deterioration of a person's state of health, in which case it is in class III; or*
- *A serious deterioration of a person's state of health or a surgical intervention, in which case it is classified as class IIb.*

Software intended to monitor physiological processes is classified as class IIa, except if it is intended for monitoring of vital physiological parameters, where the nature of variations of those parameters is such that it could result in immediate danger to the patient, in which case it is classified as class IIb.

All other software is classified as class I.

Rule 12: *All active devices intended to administer and/or remove medicinal products, body liquids or other substances to or from the body are classified as class IIa, unless this is done in a manner that is potentially hazardous, taking account of the nature of the substances involved, of the part of the body concerned and of the mode of application in which case they are classified as class IIb.*

Rule 13: *All other active devices are classified as class I.*

Special Rules

Rule 14: *All devices incorporating, as an integral part, a substance which, if used separately, can be considered to be a medicinal product, as defined in point 2 of Article 1 of Directive 2001/83/EC, including a medicinal product derived from human blood or human plasma, as defined in point 10 of Article 1 of that Directive, and that has an action ancillary to that of the devices, are classified as class III.*

Rule 15: *All devices used for contraception or prevention of the transmission of sexually transmitted diseases are classified as class IIb, unless they are implantable or long-term invasive devices, in which case they are classified as class III.*

Rule 16: *All devices intended specifically to be used for disinfecting, cleaning, rinsing, or, where appropriate, hydrating contact lenses are classified as class IIb.* *All devices intended specifically to be used for disinfecting or sterilizing medical devices are classified as class IIa, unless they are disinfecting solutions or washer-disinfectors intended specifically to be used for disinfecting invasive devices, as the end point of processing, in which case they are classified as class IIb.* *This rule does not apply to devices that are intended to clean devices other than contact lenses by means of physical action only.*
Rule 17: *Devices specifically intended for recording of diagnostic images generated by X-ray radiation are classified as class IIa.*
Rule 18: *All devices manufactured utilizing tissues or cells of human or animal origin, or their derivatives, which are non-viable or rendered non-viable, are classified as class III, unless such devices are manufactured utilizing tissues or cells of animal origin, or their derivatives, which are non-viable or rendered non-viable and are devices intended to come into contact with intact skin only.*
Rule 19: *All devices incorporating or consisting of nanomaterial are classified as:* • *Class III if they present a high or medium potential for internal exposure;* • *Class IIb if they present a low potential for internal exposure; and* • *Class IIa if they present a negligible potential for internal exposure.*
Rule 20: *All invasive devices with respect to body orifices, other than surgically invasive devices, which are intended to administer medicinal products by inhalation are classified as class IIa, unless their mode of action has an essential impact on the efficacy and safety of the administered medicinal product or they are intended to treat life-threatening conditions, in which case they are classified as class IIb.*

Rule 21: *Devices that are composed of substances or of combinations of substances that are intended to be introduced into the human body via a body orifice or applied to the skin and that are absorbed by or locally dispersed in the human body are classified as:* • *Class III if they, or their products of metabolism, are systemically absorbed by the human body in order to achieve the intended purpose;* • *Class III if they achieve their intended purpose in the stomach or lower gastrointestinal tract and they, or their products of metabolism, are systemically absorbed by the human body;* • *Class IIa if they are applied to the skin or if they are applied in the nasal or oral cavity as far as the pharynx, and achieve their intended purpose on those cavities; and* • *Class IIb in all other cases.*
Rule 22: *Active therapeutic devices with an integrated or incorporated diagnostic function which significantly determines the patient management by the device, such as closed loop systems or automated external defibrillators, are classified as class III.*

The classification rules are summarized below:

Rule 1	Non-invasive devices
Rule 2	Non-invasive devices intended for channeling or storing (Which includes cells)
Rule 3	Non-invasive devices that modify biological or chemical composition of blood, body liquids, other liquids and cells
Rule 4	Non-invasive devices in contact with injured skin or mucous membrane
Rule 5	Devices invasive in body orifices
Rule 6	Surgically invasive devices for transient use
Rule 7	Surgically invasive devices for short term use
Rule 8	Surgically invasive devices for long term use and implantable (including any device administering medicinal products, surgical mesh or spinal disc)
Rule 9	Active therapeutic devices intended to exchange or administer energy
Rule 10	Active devices for diagnosis and monitoring, emit ionizing radiation
Rule 11	Software intended to provide information which is used to take decisions with diagnosis or therapeutic purposes (from class I to class III)
Rule 12	Active devices intended to administer and/or remove medicinal products, body liquids or other substances

Rule 13	All other active devices
Rule 14	Devices incorporating a medicinal substance including human blood or plasma
Rule 15	Contraception or prevention of the transmission of sexually transmitted diseases
Rule 16	Specific disinfecting, cleaning, and rinsing devices
Rule 17	Devices specifically intended for recording of diagnostic images generated by X-ray radiation
Rule 18	Devices utilizing non-viable tissues or cells of human origin or tissues of animal or derivatives
Four New Rules Included in the MDR:	
Rule 19	Devices incorporating or consisting of nanomaterial
Rule 20	Invasive devices with respect to body orifices to administer medicinal products by inhalation
Rule 21	Substances or of combinations of substances that are intended to be introduced into the human body via a body orifice or applied to the skin and that are absorbed
Rule 22	Active therapeutic devices with an integrated or incorporated diagnostic function which significantly determines the patient management

The new regulation also comes with certain classification changes that are relevant to certain medical devices. All active implantable devices and their accessories will be considered as Class III. Any substance-based device intended to be used via a body orifice or applied on the skin may not be a class I. this means any substance-based device currently in Class I will be up-classified with the new regulation. This also means that a manufacturer with a device that will be up classified with the MDR must also comply with the stricter requirements arising from the new regulation.

5.9. COMPARISON OF CLASSIFICATION RULES BETWEEN THE MDD AND MDR

An overview of the classification rules between the MDR and MDD is presented below:

MDR		**MDD**	
Rules	**Type of Device**	**Rules**	**Type of Device**
1–4	Non-invasive devices	1–4	Non-invasive devices
5–8	Invasive devices	5–8	Invasive devices

9–13	Active devices	9–12	Active devices
14–22	Special rules	13–18	Special rules

Breakdown of each classification rule with the MDR (in comparison to the MDD).

Rules 1–4: Non-invasive devices (in comparison with MDD/ AIMD).

Rule 1	**Rule 2**	**Rule 3**	**Rule 4**
No change	Addition of "cells and tissues" to the existing language. Blood bags moved to MDR Rule 2 from Rule 18 of MDD.	Addition of human tissues and cells to blood, body liquids and other liquids. Intended for implantation or administration vs. Intended for infusion in MDD. Inclusion of organ storage solutions, IVF media into the rule which are class III	Addition of injured mucous membrane to injured skin. Replacement of 'wounds' with injuries to skin. Also covers invasive devices that come into contact with injured mucous membrane

Rules 5–8 Invasive devices (in comparison with MDD/AIMD).

Rule 5	**Rule 6**	**Rule 7**	**Rule 8**
No change-clarifications only	All devices intended specifically for direct contact with heart or central circulatory system now class III similar to devices in contact with central nervous system	All devices intended specifically for direct contact with heart or central circulatory system now class III similar to devices in contact with central nervous system	AIMD devices and accessories are class III. Breast implants and surgical meshes are class III. Total and partial joint replacements are class III. Spinal disc replacement implants or implantable devices that come into contact with spinal column are class III with some exceptions (screws, wedges, plates, and instruments)

Rule 9–13 Active devices (in comparison with MDD/ AIMD).

Rule 9	Rule 10	Rule 11	Rule 12	Rule 13
Addition of active devices intended to emit ionizing radiation for therapeutic purposes, including devices which control or monitor such devices, or which directly influence their performance, are classified as class IIb. Addition of active devices that are intended for controlling, monitoring or directly influencing the performance of active implantable devices are classified as class III	Addition of 'monitoring' to diagnosis; Active devices intended for diagnosis in clinical situations where the patient is in immediate danger as class IIb	New rule on software. Classifications range from class III-class I	Rule 11 in MDD. No change	Rule 12 in MDD. No change

Rule 14–18: special rules.

Rule 14 (Devices with Medicinal Substances)	Rule 15 (Contraceptive Devices, Devices for Prevention of Transmission of STDs)	Rule 16 (Disinfectants, Sterilizers)	Rule 17 (Devices for Recording X-Ray Diagnostic Images)	Rule 18 (Devices Utilizing Human or Animal Derivatives)
Rule 13 in MDD. Clarification that medicinal product can be derived from human blood or plasma. "Liable to act" taken out	Rule 14 in MDD. No change	Rule 15 in MDD. Addition of sterilizers to disinfectants. Disinfectants or sterilizers become IIb only if they are used for invasive devices and as the end point of processing	Rule 16 in MDD. No change-language clarified	Rule 17 in MDD. Addition of cells (to tissues). Addition of human origin cells and tissues or derivatives. The exception about contact with intact skin only, applies only to animal tissue and does not apply to human tissues or cells

Rules 19–22: special rules.

Rule 19 (Devices Incorporating or Consisting of Nanomaterials)	**Rule 20 (Body-Orifice Invasive Devices Intended to Administer Medicines by Inhalation)**	**Rule 21 (Devices Consisting of Substances and Introduced into the Body via Body Orifice or Skin and That are Absorbed by or Locally Dispersed)**	**Rule 22 (Active Therapeutic Device with an Integrated or Incorporated Diagnostic Function)**
New rule. Classifications from III to IIa based on potential for internal exposure	New rule. Classification IIa or IIb. IIb if they impact the safety and performance of the medicine or intended to treat life-threatening conditions	New rule. Classification from IIa to III based on where they are used and whether they or their products of metabolism are absorbed	New rule. Class III. Only applies if such devices significantly determine the patient management. Closed loop systems or automated external defibrillators

CHAPTER

6

Conformity Assessments

CONTENTS

For a medical device manufacturer, one of the biggest requirements is being able to demonstrate compliance to relevant regulatory authorities. The EU regulations give manufacturers the ability to choose which conformity assessment route they would like to use but this is dependent on the class of the device itself. Choosing the right conformity assessment procedure is important to ensure a manufacturer gets approval without wasting precious resources or delaying the launch of the product.

6.1. UNDERSTANDING CONFORMITY ASSESSMENT UNDER THE MDD

Article 11 of the MDD lists the choices a manufacturer has when it comes to choosing a devices conformity assessment route. The reason for the different choices is to provide flexibility as each manufacturer is different and what works well for one manufacturer might not work well for another. It should be noted that all conformity assessment procedures will require a manufacturer to have technical documentation pertaining to the device and this will include information about how it confirms to meet the ER. The notified body will them perform a technical documentation review of the submitted assessment. It is important to remember that the notified body won't issue an EC certificate without the relevant technical documentation in place and failing to obtain an EC certificate means a manufacturer is them unable to place the CE mark on the device and ultimately sell within the European Union.

There are six conformity assessment options under the MDD:

- **Annex II:** Full quality assurance system.
- **Annex III:** EC type examination.
- **Annex IV:** EC verification.
- **Annex V:** EC Declaration of conformity-production quality assurance.
- **Annex VI:** EC Declaration of conformity-product quality assurance.
- **Annex VII:** EC Declaration of conformity.

Article 11 of the MDD is highlighted below for the manufacturer:

Article 11 of the MDD 93/42/EEC: Conformity Assessment Procedures (as taken from the directive)

1. In the case of devices falling within Class III, other than devices which are custom-made or intended for clinical investigations, the manufacturer shall, in order to affix the CE marking, either:
a. Follow the procedure relating to the EC declaration of conformity set out in Annex II (full quality assurance); or
b. Follow the procedure relating to the EC type-examination set out in Annex III, coupled with:
i. The procedure relating to the EC verification set out in Annex IV; or
ii. The procedure relating to the EC declaration of conformity set out in Annex V (production quality assurance).
2. In the case of devices falling within Class IIa, other than devices which are custom-made or intended for clinical investigations, the manufacturer shall, in order to affix the CE marking, follow the procedure relating to the EC declaration of conformity set out in Annex VII, coupled with either:
a. The procedure relating to the EC verification set out in Annex IV;
b. The procedure relating to the EC declaration of conformity set out in Annex V (production quality assurance); or
c. The procedure relating to the EC declaration of conformity set out in Annex VI (product quality assurance).
Instead of applying these procedures, the manufacturer may also follow the procedure referred to in paragraph 3 (a).
3. In the case of devices falling within Class IIb, other than devices which are custom-made or intended for clinical investigations, the manufacturer shall, in order to affix the CE marking, either:
a. Follow the procedure relating to the EC declaration of conformity set out in Annex II (full quality assurance); in this case, point 4 of Annex II is not applicable; or
b. Follow the procedure relating to the EC type-examination set out in Annex III, coupled with:
i. The procedure relating to the EC verification set out in Annex IV;
ii. The procedure relating to the EC declaration of conformity set out in Annex V (production quality assurance); or
iii. The procedure relating to the EC declaration of conformity set out in Annex VI (product quality assurance).
4. The Commission shall, no later than five years from the date of implementation of this Directive, submit a report to the Council on the operation of the provisions referred to in Article 10 (1), Article 15 (1), in particular in respect of Class I and Class IIa devices, and on the operation of the provisions referred to in Annex II, Section 4.3 second and third subparagraphs and in Annex III, Section 5 second and third subparagraphs to this Directive, accompanied, if necessary, by appropriate proposals.

5. In the case of devices falling within Class I, other than devices which are custom-made or intended for clinical investigations, the manufacturer shall, in order to affix the CE marking, follow the procedure referred to in Annex VII and draw up the EC declaration of conformity required before placing the device on the market.
6. In the case of custom-made devices, the manufacturer shall follow the procedure referred to in Annex VIII and draw up the statement set out in that Annex before placing each device on the market. Member States may require that the manufacturer shall submit to the competent authority a list of such devices which have been put into service in their territory.
7. During the conformity assessment procedure for a device, the manufacturer, and/or the notified body shall take account of the results of any assessment and verification operations which, where appropriate, have been carried out in accordance with this Directive at an intermediate stage of manufacture.
8. The manufacturer may instruct his authorized representative to initiate the procedures provided for in Annexes III, IV, VII, and VIII.
9. Where the conformity assessment procedure involves the intervention of a notified body, the manufacturer, or his authorized representative, may apply to a body of his choice within the framework of the tasks for which the body has been notified.
10. The notified body may require, where duly justified, any information or data, which is necessary for establishing and maintaining the attestation of conformity in view of the chosen procedure.
11. Decisions taken by the NBs in accordance with Annexes II, III, V, and VI shall be valid for a maximum of five years and may be extended on application, made at a time agreed in the contract signed by both parties, for further periods of a maximum length of five years.
12. The records and correspondence relating to the procedures referred to in paragraphs 1 to 6 shall be in an official language of the Member State in which the procedures are carried out and/or in another Community language acceptable to the notified body.
13. By derogation from paragraphs 1 to 6, the competent authorities (CA) may authorize, on duly justified request, the placing on the market and putting into service, within the territory of the Member State concerned, of individual devices for which the procedures referred to in paragraphs 1 to 6 have not been carried out and the use of which is in the interest of protection of health.
*14. The measures designed to amend non-essential elements of this Directive, by supplementing it, relating to the means by which, in the light of technical progress and considering the intended users of the devices concerned, the information laid down in Annex I Section 13.1 may be set out, shall be adopted in accordance with the regulatory procedure with scrutiny referred to i*n Article 7(3).

The six conformity assessment options are explained in detail below:

1. Annex II: Full Quality Assurance System

The full quality assurance system conformity assessment is one of the more common assessment used and includes a full review of the manufacturers entire quality management system (QMS). This includes design, manufacturing, and final inspection. Regardless of whether a device falls into class III or not, when a manufacturer undergoes assessment, the technical documentation will still be reviewed. If the product is a class III device and annex II is followed for conformity assessment, the manufacturer will need to include a design dossier for the review. This will need to show conformity to the ER.

Annex II from the MDD is highlighted below:

Annex II of the MDD 93/42/EEC: EC Declaration of Conformity (full quality assurance system)

1. The manufacturer must ensure application of the quality system approved for the design, manufacture, and final inspection of the products concerned, as specified in Section 3 and is subject to audit as laid down in Sections 3.3 and 4 and to Community surveillance as specified in Section 5.

2. The EC declaration of conformity is the procedure whereby the manufacturer who fulfills the obligations imposed by Section 1 ensures and declares that the products concerned meet the provisions of this Directive which apply to them. The manufacturer must affix the CE marking in accordance with Article 17 and draw up a written declaration of conformity. This declaration must cover one or more medical devices manufactured, clearly identified by means of product name, product code, or other unambiguous reference, and must be kept by the manufacturer.

3. Quality system.

3.1. The manufacturer must lodge an application for assessment of his quality system with a notified body. The application must include:

- *The name and address of the manufacturer and any additional manufacturing site covered by the quality system.*
- *All the relevant information on the product or product category covered by the procedure.*
- *A written declaration that no application has been lodged with any other notified body for the same product-related quality system.*
- *The documentation on the quality system.*
- *An undertaking by the manufacturer to fulfill the obligations imposed by the quality system approved.*
- *An undertaking by the manufacturer to keep the approved quality system adequate and efficacious.*

- *An undertaking by the manufacturer to institute and keep up to date a systematic procedure to review experience gained from devices in the post-production phase, including the provisions referred to in Annex X, and to implement appropriate means to apply any necessary corrective action. This undertaking must include an obligation for the manufacturer to notify the CA of the following incidents immediately on learning of them:*

i. Any malfunction or deterioration in the characteristics and/or performance of a device, as well as any inadequacy in the instructions for use which might lead to or might have led to the death of a patient or user or to a serious deterioration in his state of health;

ii. Any technical or medical reason connected with the characteristics or performance of a device leading for the reasons referred to in subparagraph (i) to systematic recall of devices of the same type by the manufacturer.

3.2. Application of the quality system must ensure that the products conform to the provisions of this Directive which apply to them at every stage, from design to final inspection. All the elements, requirements, and provisions adopted by the manufacturer for his quality system must be documented in a systematic and orderly manner in the form of written policies and procedures such as quality programmes, quality plans, quality manuals, and quality records.

It shall include in particular the corresponding documentation, data, and records arising from the procedures referred to in point (c).

It shall include in particular an adequate description of:

a. The manufacturer's quality objectives;

b. The organization of the business and in particular:

· The organizational structures, the responsibilities of the managerial staff and their organizational authority where quality of design and manufacture of the products is concerned.

· The methods of monitoring the efficient operation of the quality system and in particular its ability to achieve the desired quality of design and of product, including control of products which fail to conform.

· Where the design, manufacture, and/or final inspection and testing of the products, or elements thereof, is carried out by a third party, the methods of monitoring the efficient operation of the quality system and in particular the type and extent of control applied to the third-party;

c. The procedures for monitoring and verifying the design of the products, including the corresponding documentation, and in particular:

· A general description of the product, including any variants planned, and its intended use(s).

· The design specifications, including the standards which will be applied and the results of the risk analysis, and also a description of the solutions adopted to fulfill the ER which apply to the products if the standards referred to in Article 5 are not applied in full.

· *The techniques used to control and verify the design and the processes and systematic measures which will be used when the products are being designed.*

· *If the device is to be connected to other device(s) in order to operate as intended, proof must be provided that it conforms to the ER when connected to any such device(s) having the characteristics specified by the manufacturer.*

· *A statement indicating whether or not the device incorporates, as an integral part, a substance or a human blood derivative referred to in section 7.4 of Annex I and the data on the tests conducted in this connection required to assess the safety, quality, and usefulness of that substance or human blood derivative, taking account of the intended purpose of the device.*

· *A statement indicating whether or not the device is manufactured utilizing tissues of animal origin as referred to in Commission Directive 2003/32/EC (1).*

· *The solutions adopted as referred to in Annex I, Chapter I, Section 2.*

· *The pre-clinical evaluation.*

· *The clinical evaluation referred to in Annex X.*

· *The draft label and, where appropriate, instructions for use.*

d. The inspection and quality assurance techniques at the manufacturing stage and in particular:

· *The processes and procedures which will be used, particularly as regards sterilization, purchasing, and the relevant documents.*

· *The product identification procedures drawn up and kept up to date from drawings, specifications or other relevant documents at every stage of manufacture;*

e. The appropriate tests and trials which will be carried out before, during, and after manufacture, the frequency with which they will take place, and the test equipment used; it must be possible to trace back the calibration of the test equipment adequately.

3.3. The notified body must audit the quality system to determine whether it meets the requirements referred to in Section 3.2. It must presume that quality systems which implement the relevant harmonized standards conform to these requirements. The assessment team must include at least one member with past experience of assessments of the technology concerned. The assessment procedure must include an assessment, on a representative basis, of the documentation of the design of the product(s) concerned, an inspection on the manufacturer's premises and, in duly substantiated cases, on the premises of the manufacturer's suppliers and/or subcontractors to inspect the manufacturing processes.

The decision is notified to the manufacturer. It must contain the conclusions of the inspection and a reasoned assessment.

3.4. The manufacturer must inform the notified body which approved the quality system of any plan for substantial changes to the quality system or the product-range covered. The notified body must assess the changes proposed and verify whether after these changes the quality system still meets the requirements referred to in Section 3.2. It must notify the manufacturer of its decision. This decision must contain the conclusions of the inspection and a reasoned assessment.

4. Examination of the design of the product.

4.1. In addition to the obligations imposed by Section 3, the manufacturer must lodge with the notified body an application for examination of the design dossier relating to the product which he plans to manufacture and which falls into the category referred to in Section 3.1.

4.2. The application must describe the design, manufacture, and performances of the product in question. It must include the documents needed to assess whether the product conforms to the requirements of this Directive, as referred to in Section 3.2 (c).

4.3. The notified body must examine the application and, if the product conforms to the relevant provisions of this Directive, issue the application with an EC design-examination certificate. The notified body may require the application to be completed by further tests or proof to allow assessment of conformity with the requirements of the Directive. The certificate must contain the conclusions of the examination, the conditions of validity, the data needed for identification of the approved design, where appropriate, and a description of the intended purpose of the product.

In the case of devices referred to in Annex I, Section 7.4, second paragraph, the notified body shall, as regards the aspects referred to in that section, consult one of the CA designated by the Member States in accordance with Directive 2001/83/EC or the EMEA before taking a decision. The opinion of the competent national authority or the EMEA must be drawn up within 210 days after receipt of valid documentation. The scientific opinion of the competent national authority or the EMEA must be included in the documentation concerning the device. The notified body will give due consideration to the views expressed in this consultation when making its decision. It will convey its final decision to the competent body concerned.

In the case of devices referred to in Annex I, Section 7.4, third paragraph, the scientific opinion of the EMEA must be included in the documentation concerning the device. The opinion of the EMEA must be drawn up within 210 days after receipt of valid documentation. The notified body will give due consideration to the opinion of the EMEA when making its decision. The notified body may not deliver the certificate if the EMEA's scientific opinion is unfavorable. It will convey its final decision to the EMEA.

In the case of devices manufactured utilizing tissues of animal origin as referred to in Directive 2003/32/EC, the notified body must follow the procedures referred to in that Directive.

4.4. Changes to the approved design must receive further approval from the notified body which issued the EC design-examination certificate wherever the changes could affect conformity with the ER of the Directive or with the conditions prescribed for use of the product. The applicant shall inform the notified body which issued the EC design-examination certificate of any such changes made to the approved design. This additional approval must take the form of a supplement to the EC design examination certificate.

5. Surveillance.

5.1. The aim of surveillance is to ensure that the manufacturer duly fulfills the obligations imposed by the approved quality system.

5.2. The manufacturer must authorize the notified body to carry out all the necessary inspections and supply it with all relevant information, in particular:

- *The documentation on the quality system.*
- *The data stipulated in the part of the quality system relating to design, such as the results of analyses, calculations, tests, the solutions adopted as referred to in Annex I, Chapter I, Section 2, pre-clinical, and clinical evaluation, post-market clinical follow-up plan and the results of the post-market clinical follow-up, if applicable, etc.*
- *The data stipulated in the part of the quality system relating to manufacture, such as inspection reports and test data, calibration data, qualification reports of the personnel concerned, etc.*

5.3. The notified body must periodically carry out appropriate inspections and assessments to make sure that the manufacturer applies the approved quality system and must supply the manufacturer with an assessment report.

5.4. In addition, the notified body may pay unannounced visits to the manufacturer. At the time of such visits, the notified body may, where necessary, carry out or ask for tests in order to check that the quality system is working properly. It must provide the manufacturer with an inspection report and, if a test has been carried out, with a test report.

6. Administrative provisions.

6.1. The manufacturer or his authorized representative must, for a period ending at least five years, and in the case of implantable devices at least 15 years, after the last product has been manufactured, keep at the disposal of the national authorities:

- *The declaration of conformity.*
- *The documentation referred to in the fourth indent of Section 3.1 and in particular the documentation, data, and records referred to in the second paragraph of Section 3.2.*
- *The changes referred to in Section 3.4.*
- *The documentation referred to in Section 4.2, and*
- *The decisions and reports from the notified body as referred to in Sections 3.3, 4.3, 4.4, 5.3 and 5.4.*

7. Application to devices in Classes IIa and IIb.
7.1. In line with Article 11(2) and (3), this Annex may apply to products in Classes IIa and IIb. Section 4, however, does not apply.
7.2. For devices in Class IIa the notified body shall assess, as part of the assessment in Section 3.3, the technical documentation as described in Section 3.2(c) for at least one representative sample for each device subcategory for compliance with the provisions of this Directive.
7.3. For devices in Class IIb the notified body shall assess, as part of the assessment in Section 3.3, the technical documentation as described in Section 3.2(c) for at least one representative sample for each generic device group for compliance with the provisions of this Directive.
7.4. In choosing representative sample(s) the notified body shall take into account the novelty of the technology, similarities in design, technology, manufacturing, and sterilization methods, the intended use and the results of any previous relevant assessments (e.g., with regard to physical, chemical or biological properties) that have been carried out in accordance with this Directive. The notified body shall document and keep available to the competent authority its rationale for the sample(s) taken.
7.5. Further samples shall be assessed by the notified body as part of the surveillance assessment referred to in Section 5.
8. Application to the devices referred to Article 1(4a).
*Upon completing the manufacture of each batch of devices referred to in Article 1(4a), the manufacturer shall inform the notified body of the release of the batch of devices and send to it the official certificate concerning the release of the batch of human blood derivative used in the device, issued by a State laboratory or a laboratory designated for that purpose by a Member State in accordance with Article 114(2) of Directive 2001/8*3/EC.

2. Annex III: EC Type Examination

The EC type examination conformity assessment, the notified body will examine a certain representative sample of the device to show conformity to the ER. These assessments look similar to that of the design dossier assessments. In this type of assessment, the notified body will perform an off-site, in-depth review of the technical requirements.

Annex III from the MDD is highlighted below:

Annex III of the MDD 93/42/EEC: EC Type Examination

1. EC type-examination is the procedure whereby a notified body ascertains and certifies that a representative sample of the production covered fulfills the relevant provisions of this Directive.

2. The application includes:

· The name and address of the manufacturer and the name and address of the authorized representative if the application is lodged by the representative.

· The documentation described in Section 3 needed to assess the conformity of the representative sample of the production in question, hereinafter referred to as the 'type,' with the requirements of this Directive. The applicant must make a 'type' available to the notified body. The notified body may request other samples as necessary.

· A written declaration that no application has been lodged with any other notified body for the same type.

3. The documentation must allow an understanding of the design, the manufacture and the performances of the product and must contain the following items in particular:

· A general description of the type, including any variants planned, and its intended use(s).

· Design drawings, methods of manufacture envisaged, in particular as regards sterilization, and diagrams of components, sub-assemblies, circuits, etc.

· The descriptions and explanations necessary to understand the above-mentioned drawings and diagrams and the operation of the product.

· A list of the standards referred to in Article 5, applied in full or in part, and descriptions of the solutions adopted to meet the ER if the standards referred to in Article 5 have not been applied in full.

· The results of the design calculations, risk analysis, investigations, technical tests, etc. carried out.

· A statement indicating whether or not the device incorporates, as an integral part, a substance, or human blood derivative, referred to in Section 7.4 of Annex I, and the data on the tests conducted in this connection which are required to assess the safety, quality, and usefulness of that substance, or human blood derivative, taking account of the intended purpose of the device.

· A statement indicating whether or not the device is manufactured utilizing tissues of animal origin as referred to in Directive 2003/32/EC.

· The solutions adopted as referred to in Annex I, Chapter I, Section 2.

· The pre-clinical evaluation.

· The clinical evaluation referred to in Annex X.

· The draft label and, where appropriate, instructions for use.

4. The notified body must:

4.1. Examine and assess the documentation and verify that the type has been manufactured in conformity with that documentation; it must also record the items designed in conformity with the applicable provisions of the standards referred to in Article 5, as well as the items not designed on the basis of the relevant provisions of the above-mentioned standards;

4.2. Carry out or arrange for the appropriate inspections and the tests necessary to verify whether the solutions adopted by the manufacturer meet the ER of this Directive if the standards referred to in Article 5 have not been applied; if the device is to be connected to other device(s) in order to operate as intended, proof must be provided that it conforms to the ER when connected to any such device(s) having the characteristics specified by the manufacturer;
4.3. Carry out or arrange for the appropriate inspections and the tests necessary to verify whether, if the manufacturer has chosen to apply the relevant standards, these have actually been applied;
4.4. Agree with the applicant on the place where the necessary inspections and tests will be carried out.
5. If the type conforms to the provisions of this Directive, the notified body issues the applicant with an EC type-examination certificate. The certificate must contain the name and address of the manufacturer, the conclusions of the inspection, the conditions of validity, and the data needed for identification of the type approved. The relevant parts of the documentation must be annexed to the certificate and a copy kept by the notified body.
In the case of devices referred to in Annex I, Section 7.4, second paragraph, the notified body shall, as regards the aspects referred to in that section, consult one of the authorities designated by the Member States in accordance with Directive 2001/83/EC or the EMEA before taking a decision. The opinion of the competent national authority or the EMEA must be drawn up within 210 days after receipt of valid documentation. The scientific opinion of the competent national authority or the EMEA must be included in the documentation concerning the device. The notified body will give due consideration to the views expressed in this consultation when making its decision. It will convey its final decision to the competent body concerned.
In the case of devices referred to in Annex I, Section 7.4, third paragraph, the scientific opinion of the EMEA must be included in the documentation concerning the device. The opinion of the EMEA must be drawn up within 210 days after receipt of valid documentation. The notified body will give due consideration to the opinion of the EMEA when making its decision. The notified body may not deliver the certificate if the EMEA's scientific opinion is unfavorable. It will convey its final decision to the EMEA.
In the case of devices manufactured utilizing tissues of animal origin as referred to in Directive 2003/32/EC, the notified body must follow the procedures referred to in that Directive.

6. The applicant must inform the notified body which issued the EC type-examination certificate of any significant change made to the approved product.

Changes to the approved product must receive further approval from the notified body which issued the EC type-examination certificate wherever the changes may affect conformity with the ER or with the conditions prescribed for use of the product. This new approval must, where appropriate, take the form of a supplement to the initial EC type-examination certificate.

7. Administrative provisions.

7.2. Other NBs may obtain a copy of the EC type-examination certificates and/or the supplements thereto. The Annexes to the certificates must be made available to other NBs on reasoned application, after the manufacturer has been informed.

7.3. The manufacturer or his authorized representative must keep with the technical documentation copies of EC type-examination certificates and their additions for a period ending at least five years after the last device has been manufactured. In the case of implantable devices, the period shall be at least 15 years after the last product has been manufactured.

3. Annex IV: EC Verification

In an EC verification type of conformity assessment, the notified body will examine and test the product to verify its conformity. The notified body will either test or assess each individual product, or they will use a statistical sampling method to make sure that they test and assess an adequate sample size to determine conformity. If a manufacturer passes, then the notified body will certify each product or each batch. The certificate issued will only apply to that product or batch that is examined. It should be noted that while this assessment does not have QMS requirements, it does have requirements for the manufacturer's post-market surveillance system. It also requires a manufacturer to provide technical documentation for those devices that demonstrate conformity to the ER.

Annex IV from the MDD is highlighted below:

Annex IV of the MDD 93/42/EEC: EC Verification

1. EC verification is the procedure whereby the manufacturer or his authorized representative ensures and declares that the products which have been subject to the procedure set out in Section 4 conform to the type described in the EC type-examination certificate and meet the requirements of this Directive which apply to them.

2. The manufacturer must take all the measures necessary to ensure that the manufacturing process produces products which conform to the type described in the EC type-examination certificate and to the requirements of the Directive which apply to them. Before the start of manufacture, the manufacturer must prepare documents defining the manufacturing process, in particular as regards sterilization where necessary, together with all the routine, pre-established provisions to be implemented to ensure homogeneous production and, where appropriate, conformity of the products with the type described in the EC type-examination certificate and with the requirements of this Directive which apply to them. The manufacturer must affix the CE marking in accordance with Article 17 and draw up a declaration of conformity.

In addition, for products placed on the market in sterile condition, and only for those aspects of the manufacturing process designed to secure and maintain sterility, the manufacturer must apply the provisions of Annex V, Sections 3 and 4.

3. The manufacturer must undertake to institute and keep up to date a systematic procedure to review experience gained from devices in the post production phase, including the provisions referred to in Annex X, and to implement appropriate means to apply any necessary corrective action. This undertaking must include an obligation for the manufacturer to notify the CA of the following incidents immediately on learning of them:

i. Any malfunction or deterioration in the characteristics and/or performance of a device, as well as any inadequacy in the labeling the instructions for use which might lead to or might have led to the death of a patient or user or to a serious deterioration in his state of health;

ii. Any technical or medical reason connected with the characteristics or performance of a device for the reasons referred to in subparagraph (i) leading to systematic recall of devices of the same type by the manufacturer.

4. The notified body must carry out the appropriate examinations and tests in order to verify the conformity of the product with the requirements of the Directive either by examining and testing every product as specified in Section 5 or by examining and testing products on a statistical basis as specified in Section 6, as the manufacturer decides. The aforementioned checks do not apply to those aspects of the manufacturing process designed to secure sterility.

5. Verification by examination and testing of every product.

5.1. Every product is examined individually and the appropriate tests defined in the relevant standard(s) referred to in Article 5 or equivalent tests must be carried out in order to verify, where appropriate, the conformity of the products with the EC type described in the type-examination certificate and with the requirements of the Directive which apply to them.

5.2. The notified body must affix, or have affixed its identification number to each approved product and must draw up a written certificate of conformity relating to the tests carried out.

6. Statistical verification.

6.1. The manufacturer must present the manufactured products in the form of homogeneous batches.

6.2. A random sample is taken from each batch. The products which make up the sample are examined individually and the appropriate tests defined in the relevant standard(s) referred to in Article 5 or equivalent tests must be carried out to verify, where appropriate, the conformity of the products with the type described in the EC type-examination certificate and with the requirements of the Directive which apply to them in order to determine whether to accept or reject the batch.

6.3. Statistical control of products will be based on attributes and/or variables, entailing sampling schemes with operational characteristics which ensure a high level of safety and performance according to the state of the art. The sampling schemes will be established by the harmonized standards referred to in Article 5, taking account of the specific nature of the product categories in question.

6.4. If the batch is accepted, the notified body affixes or has affixed its identification number to each product and draws up a written certificate of conformity relating to the tests carried out. All products in the batch may be put on the market except any in the sample which failed to conform.

If a batch is rejected, the competent notified body must take appropriate measures to prevent the batch from being placed on the market. In the event of frequent rejection of batches, the notified body may suspend the statistical verification. The manufacturer may, on the responsibility of the notified body, affix the notified body's identification number during the manufacturing process.

7. Administrative provisions.

The manufacturer or his authorized representative must, for a period ending at least five years, and in the case of implantable devices at least 15 years, after the last product has been manufactured, make available to the national authorities:

- *The declaration of conformity.*
- *The documentation referred to in Section 2.*
- *The certificates referred to in Sections 5.2 and 6.4.*
- *Where appropriate, the type-examination certificate referred to in Annex III.*

8. Application to devices in Class IIa.

In line with Article 11 (2), this Annex may apply to products in Class IIa, subject to the following:

8.1. In derogation from Sections 1 and 2, by virtue of the declaration of conformity the manufacturer ensures and declares that the products in Class IIa are manufactured in conformity with the technical documentation referred to in Section 3 of Annex VII and meet the requirements of this Directive which apply to them;

8.2. In derogation from Sections 1, 2, 5 and 6, the verifications conducted by the notified body are intended to confirm the conformity of the products in Class IIa with the technical documentation referred to in Section 3 of Annex VII.

9. Application to devices referred to in Article 1(4a).
*In the case of section 5, upon completing the manufacture of each batch of devices referred to in Article 1(4a), and in the case of verification under section 6, the manufacturer shall inform the notified body of the release of this batch of devices and send to it the official certificate concerning the release of the batch of human blood derivative used in the device issued by a State laboratory or a laboratory designated for that purpose by a Member State in accordance with Article 114(2) of Directive 2001/83/*EC.

4. Annex V: EC Declaration of Conformity-Production Quality Assurance

It is important to understand that while Annex V shares some similarities with Annex II when it comes to QMS requirements. It doesn't include design requirements. This means it requires you to have a QMS for your manufacturing and a final inspection for your devices, but it does not require you to include a design system. This means there may not be design history file (DHF) in your technical documentation. It should be noted that technical documentation is still a requirement and, in some cases, Annex V will be combined with Annex III (EC Type Examination).

Annex V from the MDD is highlighted below:

Annex V of the MDD 93/42/EEC: EC Declaration of Conformity Production Quality Assurance

1. The manufacturer must ensure application of the quality system approved for the manufacture of the products concerned and carry out the final inspection, as specified in Section 3, and is subject to the Community surveillance referred to in Section 4.
2. The EC declaration of conformity is the part of the procedure whereby the manufacturer who fulfills the obligations imposed by Section 1 ensures and declares that the products concerned conform to the type described in the EC type-examination certificate and meet the provisions of this Directive which apply to them.
The manufacturer must affix the CE marking in accordance with Article 17 and draw up a written declaration of conformity. This declaration must cover one or more medical devices manufactured, clearly identified by means of product name, product code or other unambiguous reference, and must be kept by the manufacturer.
3. Quality system.
3.1. The manufacturer must lodge an application for assessment of his quality system with a notified body.

The application must include:

- *The name and address of the manufacturer.*
- *All the relevant information on the product or product category covered by the procedure.*
- *A written declaration that no application has been lodged with any other notified body for the same products.*
- *The documentation on the quality system.*
- *An undertaking to fulfill the obligations imposed by the quality system is approved.*
- *An undertaking to maintain the practicability and effectiveness of the approved quality system.*
- *Where appropriate, the technical documentation on the types approved and a copy of the EC type-examination certificates.*
- *An undertaking by the manufacturer to institute and keep up to date a systematic procedure to review experience gained from devices in the post-production phase, including the provisions referred to in Annex X, and to implement appropriate means to apply any necessary corrective action. This undertaking must include an obligation for the manufacturer to notify the CA of the following incidents immediately on learning of them:*

i. Any malfunction or deterioration in the characteristics and/or performance of a device, as well as any inadequacy in the labeling or the instructions for use which might lead to or might have led to the death of a patient or user or to a serious deterioration in his state of health;

ii. Any technical or medical reason connected with the characteristics or performance of a device for the reasons referred to in subparagraph (i) above leading to a systematic recall of devices of the same type by the manufacturer.

3.2. Application of the quality system must ensure that the products conform to the type described in the EC type-examination certificate.

All the elements, requirements, and provisions adopted by the manufacturer for his quality system must be documented in a systematic and orderly manner in the form of written policy statements and procedures. This quality system documentation must permit uniform interpretation of the quality policy and procedures such as quality programmes, plans, manuals, and records.

It must include in particular an adequate description of:

a. The manufacturer's quality objectives;

b. The organization of the business and in particular:

- *The organizational structures, the responsibilities of the managerial staff and their organizational authority where manufacture of the products is concerned.*
- *The methods of monitoring the efficient operation of the quality system and in particular its ability to achieve the desired quality of product, including control of products which fail to conform.*

- *Where the manufacture and/or final inspection and testing of the products, or elements thereof, are carried out by a third party, the methods of monitoring the efficient operation of the quality system and in particular the type and extent of control applied to the third-party;*

c. The inspection and quality assurance techniques at the manufacturing stage and in particular:

- *The processes and procedures which will be used, particularly as regards sterilization, purchasing, and the relevant documents.*
- *The product identification procedures drawn up and kept up to date from drawings, specifications or other relevant documents at every stage of manufacture;*

d. The appropriate tests and trials to be carried out before, during, and after manufacture, the frequency with which they will take place, and the test equipment used; it must be possible adequately to trace back the calibration of the test equipment.

3.3. The notified body must audit the quality system to determine whether it meets the requirements referred to in Section 3.2. It must presume that quality systems which implement the relevant harmonized standards conform to these requirements.

The assessment team must include at least one member with past experience of assessments of the technology concerned. The assessment procedure must include an inspection on the manufacturer's premises and, in duly substantiated cases, on the premises of the manufacturer's suppliers to inspect the manufacturing processes.

The decision must be notified to the manufacturer after the final inspection and contain the conclusions of the inspection and a reasoned assessment.

3.4. The manufacturer must inform the notified body which approved the quality system of any plan for substantial changes to the quality system.

The notified body must assess the changes proposed and verify whether after these changes the quality system still meets the requirements referred to in Section 3.2.

After the above-mentioned information has been received, the decision is notified to the manufacturer. It must contain the conclusions of the inspection and a reasoned assessment.

4. Surveillance.

4.1. The aim of surveillance is to ensure that the manufacturer duly fulfills the obligations imposed by the approved quality system.

4.2. The manufacturer authorizes the notified body to carry out all the necessary inspections and must supply it with all relevant information, in particular:

- *The documentation on the quality system.*
- *The technical documentation.*
- *The data stipulated in the part of the quality system relating to manufacture, such as inspection reports and test data, calibration data, qualification reports of the personnel concerned, etc.*

4.3. The notified body must periodically carry out appropriate inspections and assessments to make sure that the manufacturer applies the approved quality system and supply the manufacturer with an assessment report.

4.4. In addition, the notified body may pay unannounced visits to the manufacturer. At the time of such visits, the notified body may, where necessary, carry out or ask for tests in order to check that the quality system is working properly. It must provide the manufacturer with an inspection report and, if a test has been carried out, with a test report.

5. Administrative provisions.

5.1. The manufacturer or his authorized representative must, for a period ending at least five years, and in the case of implantable devices at least 15 years, after the last product has been manufactured, make available to the national authorities:

- *The declaration of conformity.*
- *The documentation referred to in the fourth indent of Section 3.1.*
- *The changes referred to in Section 3.4.*
- *The documentation referred to in the seventh indent of Section 3.1.*
- *The decisions and reports from the notified body as referred to in Sections 4.3 and 4.4.*
- *Where appropriate, the type-examination certificate referred to in Annex III.*

6. Application to devices in Class IIa.

In line with Article 11(2), this Annex may apply to products in Class IIa, subject to the following:

6.1. By way of derogation from Sections 2, 3.1 and 3.2, by virtue of the declaration of conformity the manufacturer ensures and declares that the products in Class IIa are manufactured in conformity with the technical documentation referred to in Section 3 of Annex VII and meet the requirements of this Directive which apply to them.

6.2. For devices in Class IIa the notified body shall assess, as part of the assessment in Section 3.3, the technical documentation as described in Section 3 of Annex VII for at least one representative sample for each device subcategory for compliance with the provisions of this Directive.

6.3. In choosing representative sample(s) the notified body shall take into account the novelty of the technology, similarities in design, technology, manufacturing, and sterilization methods, the intended use and the results of any previous relevant assessments (e.g., with regard to physical, chemical or biological properties) that have been carried out in accordance with this Directive. The notified body shall document and keep available to the competent authority its rationale for the sample(s) taken.

6.4. Further samples shall be assessed by the notified body as part of the surveillance assessment referred to in Section 4.3.

7. Application to devices referred to in Article 1(4a).
Upon completing the manufacture of each batch of devices referred to in Article 1(4a), the manufacturer shall inform the notified body of the release of the batch of devices and send to it the official certificate concerning the release of the batch of human blood derivative used in the device issued by a State laboratory or a laboratory designated for that purpose by a Member State in accordance with Article 114(2) of Directive 2001/83/EC.

5. Annex VI: EC Declaration of Conformity-Product Quality Assurance

Annex VI requires a quality system for final inspection and testing, but a manufacturer won't need a QMS that includes design or manufacturing. For that reason, this type of assessment is typically combined with Annex III for class IIb devices. This can, however, result in a lengthy review process for the device. This assessment may also not be feasible if the devices require a special type of manufacturing process.

Annex VI from the MDD is highlighted below:

Annex VI of the MDD 93/42/EEC: EC Declaration of Conformity Product Quality Assurance

1. The manufacturer must ensure application of the quality system approved for the final inspection and testing of the product, as specified in Section 3 and must be subject to the surveillance referred to in Section 4.
In addition, for products placed on the market in sterile condition, and only for those aspects of the manufacturing process designed to secure and maintain sterility, the manufacturer must apply the provisions of Annex V, Sections 3 and 4.
2. The EC declaration of conformity is the part of the procedure whereby the manufacturer who fulfills the obligations imposed by Section 1 ensures and declares that the products concerned conform to the type described in the EC type-examination certificate and meet the provisions of this Directive which apply to them.
The manufacturer affixes the CE marking in accordance with Article 17 and draws up a written declaration of conformity. This declaration must cover one or more medical devices manufactured, clearly identified by means of product name, product code or other unambiguous reference, and are kept by the manufacturer. The CE marking must be accompanied by the identification number of the notified body which performs the tasks referred to in this Annex.

3. Quality system.

3.1. The manufacturer lodges an application for assessment of his quality system with a notified body.

The application must include:

- *The name and address of the manufacturer.*
- *All the relevant information on the product or product category covered by the procedure.*
- *A written declaration specifying that no application has been lodged with any other notified body for the same products.*
- *The documentation on the quality system.*
- *An undertaking by the manufacturer to fulfill the obligations imposed by the quality system approved.*
- *An undertaking by the manufacturer to keep the approved quality system adequate and efficacious.*
- *Where appropriate, the technical documentation on the types approved and a copy of the EC type-examination certificates.*
- *An undertaking by the manufacturer to institute and keep up to date a systematic procedure to review experience gained from devices in the post-production phase, including the provisions referred to in Annex X, and to implement appropriate means to apply any necessary corrective action. This undertaking must include an obligation for the manufacturer to notify the CA of the following incidents immediately on learning of them:*

i. Any malfunction or deterioration in the characteristics and/or performance of a device, as well as any inadequacy in the labeling or the instructions for use which might lead to or might have led to the death of a patient or user or to a serious deterioration in his state of health;

ii. Any technical or medical reason connected with the characteristics or the performance of a device for the reasons referred to in subparagraph.

iii. Leading to a systematic recall of devices of the same type by the manufacturer.

3.2. Under the quality system, each product or a representative sample of each batch is examined and the appropriate tests defined in the relevant standard(s) referred to in Article 5 or equivalent tests are carried out to ensure that the products conform to the type described in the EC type-examination certificate and fulfil the provisions of this Directive which apply to them. All the elements, requirements, and provisions adopted by the manufacturer must be documented in a systematic and orderly manner in the form of written measures, procedures, and instructions. This quality system documentation must permit uniform interpretation of the quality programmes, quality plans, quality manuals, and quality records.

It must include in particular an adequate description of:

- *The quality objectives and the organizational structure, responsibilities and powers of the managerial staff with regard to product quality.*

- *The examinations and tests that will be carried out after manufacture; it must be possible to trace back the calibration of the test equipment adequately.*
- *The methods of monitoring the efficient operation of the quality system.*
- *The quality records, such as reports concerning inspections, tests, calibration and the qualifications of the staff concerned, etc.*
- *Where the final inspection and testing of the products, or elements thereof, are carried out by a third party, the methods of monitoring the efficient operation of the quality system and in particular the type and extent of control applied to the third party.*

The aforementioned checks do not apply to those aspects of the manufacturing process designed to secure sterility.

3.3. The notified body audits the quality system to determine whether it meets the requirements referred to in section 3.2. It must presume that quality systems which implement the relevant harmonized standards conform to these requirements.

The assessment team must include at least one member with past experience of assessments of the technology concerned. The assessment procedure must include an inspection on the manufacturer's premises and, in duly substantiated cases, on the premises of the manufacturer's suppliers to inspect the manufacturing processes.

The decision must be notified to the manufacturer. It must contain the conclusions of the inspection and a reasoned assessment.

3.4. The manufacturer must inform the notified body which approved the quality system of any plan for substantial changes to the quality system.

The notified body must assess the changes proposed and verify whether after these changes the quality system will still meet the requirements referred to in Section 3.2.

After receiving the above-mentioned information, it must notify the manufacturer of its decision. This decision must contain the conclusions of the inspection and a reasoned assessment.

4. Surveillance.

4.1. The aim of surveillance is to ensure that the manufacturer duly fulfills the obligations imposed by the approved quality system.

4.2. The manufacturer must allow the notified body access for inspection purposes to the inspection, testing, and storage locations and supply it with all relevant information, in particular:

- *The documentation on the quality system.*
- *The technical documentation.*
- *The quality records, such as inspection reports, test data, calibration data, qualification reports of the staff concerned, etc.*

4.3. The notified body must periodically carry out appropriate inspections and assessments to make sure that the manufacturer applies the quality system and must supply the manufacturer with an assessment report.

4.4. In addition, the notified body may pay unannounced visits to the manufacturer. At the time of such visits, the notified body may, where necessary, carry out or ask for tests in order to check that the quality system is working properly and that the production conforms to the requirements of the Directive which apply to it. To this end, an adequate sample of the final products, taken on site by the notified body, must be examined and the appropriate tests defined in the relevant standard(s) referred to in Article 5 or equivalent tests must be carried out. Where one or more of the samples fails to conform, the notified body must take the appropriate measures.

It must provide the manufacturer with an inspection report and, if a test has-been carried out, with a test report.

5. Administrative provisions.

5.1. The manufacturer or his authorized representative must, for a period ending at least five years, and in the case of implantable devices at least 15 years, after the last product has been manufactured, make available to the national authorities:

- *The declaration of conformity.*
- *The documentation referred to in the seventh indent of Section 3.1.*
- *The changes referred to in Section 3.4.*
- *The decisions and reports from the notified body as referred to in the final indent of Section 3.4 and in Sections 4.3 and 4.4.*
- *Where appropriate, the certificate of conformity referred to in Annex III.*

6. Application to devices in Class IIa.

In line with Article 11(2), this Annex may apply to products in Class IIa, subject to the following:

6.1. By way of derogation from Sections 2, 3.1 and 3.2, by virtue of the declaration of conformity the manufacturer ensures and declares that the products in Class IIa are manufactured in conformity with the technical documentation referred to in Section 3 of Annex VII and meet the requirements of this Directive which apply to them.

6.2. For devices in Class IIa the notified body shall assess, as part of the assessment in Section 3.3, the technical documentation as described in Section 3 of Annex VII for at least one representative sample for each device subcategory for compliance with the provisions of this Directive.

6.3. In choosing representative sample(s) the notified body shall take into account the novelty of the technology, similarities in design, technology, manufacturing, and sterilization methods, the intended use and the results of any previous relevant assessments (e.g., with regard to physical, chemical or biological properties) that have been carried out in accordance with this directive. The notified body shall document and keep available to the competent authority its rationale for the sample(s) taken.

6.4. Further samples shall be assessed by the notified body as part of the surveillance assessment referred to in Section 4.3.

6. Annex VII: EC Declaration of Conformity

Annex VII requires you to have a systematic procedure in place to collect post market information about the devices in use, when in the field/on the market. During the assessment, you will be required to provide technical documentation demonstrate conformity with ER.

Annex VII from the MDD is highlighted below:

Annex VII of the MDD 93/42/EEC: EC Declaration of Conformity

1. The EC declaration of conformity is the procedure whereby the manufacturer or his authorized representative who fulfills the obligations imposed by Section 2 and, in the case of products placed on the market in a sterile condition and devices with a measuring function, the obligations imposed by Section 5 ensures and declares that the products concerned meet the provisions of this Directive which apply to them.

2. The manufacturer must prepare the technical documentation described in Section 3. The manufacturer or his authorized representative must make this documentation, including the declaration of conformity, available to the national authorities for inspection purposes for a period ending at least five years after the last product has been manufactured. In the case of implantable devices the period shall be at least 15 years after the last product has been manufactured.

3. The technical documentation must allow assessment of the conformity of the product with the requirements of the Directive. It must include in particular:

- *A general description of the product, including any variants planned and its intended use(s).*
- *Design drawings, methods of manufacture envisaged and diagrams of components, sub-assemblies, circuits, etc.*
- *The descriptions and explanations necessary to understand the above-mentioned drawings and diagrams and the operations of the product.*
- *The results of the risk analysis and a list of the standards referred to in Article 5, applied in full or in part, and descriptions of the solutions adopted to meet the ER of the Directive if the standards referred to in Article 5 have not been applied in full.*
- *In the case of products placed on the market in a sterile condition, description of the methods used and the validation report.*
- *The results of the design calculations and of the inspections carried out, etc.; if the device is to be connected to other device(s) in order to operate as intended, proof must be provided that it conforms to the ER when connected to any such device(s) having the characteristics specified by the manufacturer.*
- *The solutions adopted as referred to in Annex I, Chapter I, Section 2.*
- *The pre-clinical evaluation.*

- *The clinical evaluation in accordance with Annex X.*
- *The label and instructions for use.*

4. The manufacturer shall institute and keep up to date a systematic procedure to review experience gained from devices in the post-production phase, including the provisions referred to in Annex X, and to implement appropriate means to apply any necessary corrective actions, taking account of the nature and risks in relation to the product. He shall notify the CA of the following incidents immediately on learning of them:

i. Any malfunction or deterioration in the characteristics and/or performance of a device, as well as any inadequacy in the labeling the instructions for use which might lead to or might have led to the death of a patient or user or to a serious deterioration in his state of health;

ii. Any technical or medical reason connected with the characteristics on the performance of a device for the reasons referred to in subparagraph (i) leading to systematic recall of devices of the same type by the manufacturer.

5. With products placed on the market in sterile condition and Class I devices with a measuring function, the manufacturer must observe not only the provisions laid down in this Annex but also one of the procedures referred to in Annex II, IV, V or VI. Application of the above-mentioned Annexes and the intervention by the notified body is limited to:

- *In the case of products placed on the market in sterile condition, only the aspects of manufacture concerned with securing and maintaining sterile conditions.*
- *In the case of devices with a measuring function, only the aspects of manufacture concerned with the conformity of the products with the metrological requirements.*

Section 6.1 of this Annex is applicable.

6. Application to devices in Class IIa.

In line with Article 11 (2), this Annex may apply to products in Class IIa, subject to the following derogation:

6.1. Where this Annex is applied in conjunction with the procedure referred to in Annex IV, V or VI, the declaration of conformity referred to in the above-mentioned Annexes forms a single declaration. As regards the declaration based on this Annex, the manufacturer must ensure and declare that the product design meets the provisions of this Directive which apply to it.

A manufacturer must remember that the classification of the device dictates the appropriate conformity assessment procedure that should be used. These are further explained below as designated by the MDD (Table 6.1).

Table 6.1. Overview of Device Classification and Appropriate Conformity Assessment Procedures

Conformity Assessment Procedures	**Classes**					
Annexes	**I**	**I (Sterile)**	**I (Measuring)**	**IIA**	**IIB**	**III**
II (+sect 4)						X
II (-sect 4)		X	X	X	X	
III					X	X
IV		X	X	X	X	X
V		X	X	X	X	X
VI		X	X	X	X	
VII	X	X	X	X		

6.1.1. Class I Medical Device Conformity Assessment Routes

For a class I medical device, the manufacturer is responsible for ensuring that his product complies with all the relevant Essential Requirements (ER) of the Directive.

It is the manufacturer's responsibility to draw up a written statement indicating this, also known as a self-declaration. Class I medical devices without a measuring function and which are supplied in a non-sterile condition does not require the involvement of a Notified Body.

It should also be noted that conformity to the International and European ISO 13485 standard is voluntary.

Manufacturers of sterile products and devices with a measuring function must apply to a Notified Body for certification of those manufacturing aspects related to sterility or metrology.

Once the manufacturer is satisfied that his product meets all the relevant ER, the manufacturer, or his European

Authorized Representative can register with the Competent Authority (CA), followed by affixing a CE mark on the product. The device is then ready to be placed on the European market (Figure 6.1).

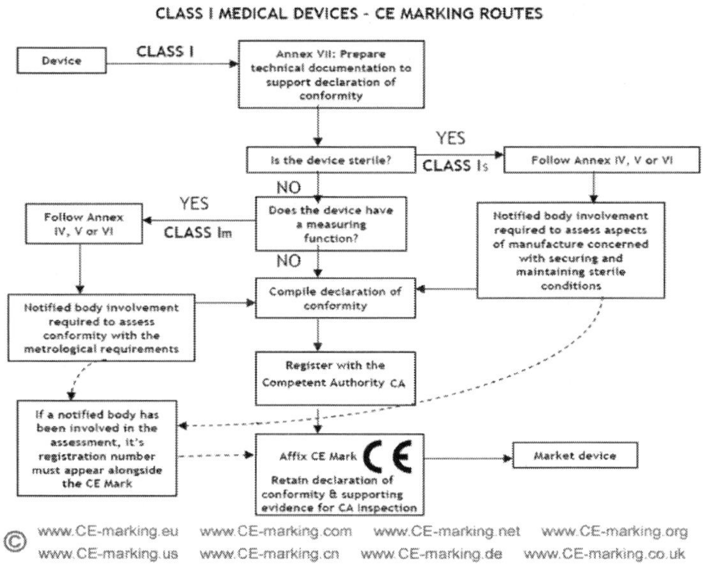

Figure 6.1. Class I medical devices CE marking routes.

Source: http://www.ce-marking.com/medical-devices-class-i.html.

Simplified Steps for Class I Medical Device Compliance	
1.	*Classification: ensure the device is a class I medical device;*
2.	*Choose relevant conformity assessment route;*
3.	*Compile the technical file;*
4.	*Draw up the declaration of conformity;*
5.	*Appoint an authorized representative;*
6.	*Register with the competent authority;*
7.	*Affix CE marking and market the products;*
8.	*Maintain vigilance monitoring and post market surveillance.*

6.1.2. Class IIa Medical Device Conformity Assessment Routes

For class IIa medical devices (other than devices which are custom-made or intended for clinical investigations), the manufacturer shall either:

1. Follow the procedure relating to the EC declaration of conformity set out in Annex II (full quality assurance); in this case, point 4 of Annex II is not applicable; or
2. Follow the procedure relating to the EC declaration of conformity set out in Annex VII, coupled with either:

i. The procedure relating to the EC verification set out in Annex IV;

ii. The procedure relating to the EC declaration of conformity set out in Annex V (production quality assurance); or

iii. The procedure relating to the EC declaration of conformity set out in Annex VI (product quality assurance).

For Class IIa products, the declaration must be backed up in all cases with conformity assessment by Notified Body and this assessment may consist of:

1. Examination and testing of each product or homogenous batch of products (Annex IV);
2. Audit of the production quality assurance system (Annex V :) ISO 13485 (excluding Design);
3. Audit of final inspection and testing (Annex VI :) ISO 13485 (excluding Design and Manufacture) or
4. Audit of the full quality assurance system (Annex II) ISO 13485.

This is decided upon by the manufacturer. Once the manufacturer has received certification from the Notified Body they may CE mark the product and place them on the market (Figure 6.2).

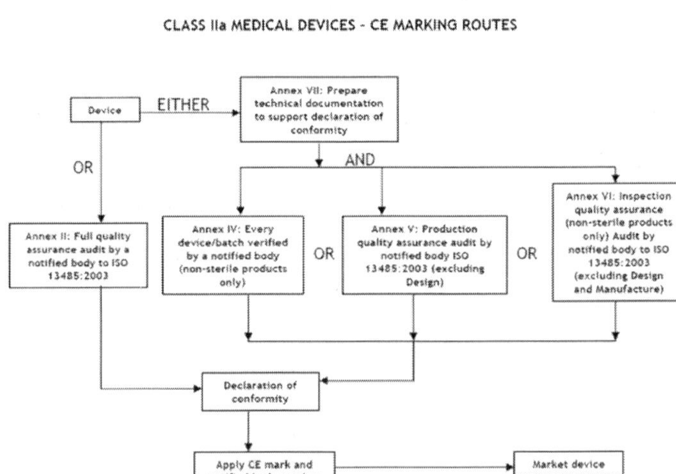

Figure 6.2. Class IIa medical devices CE marking routes.

Simplified Steps for Class IIa Medical Device Compliance	
1.	*Classification: ensure the device is a Class IIa medical device;*
2.	*Choose conformity assessment route;*
3.	*Compile the technical file;*
4.	*Obtain certification from a notified body;*
5.	*Draw up the declaration of conformity;*
6.	*Appoint an authorized representative;*
7.	*Hold the tech files for inspection by the competent authority;*
8.	*Affix CE marking and market the products;*
9.	*Maintain vigilance monitoring and post market surveillance.*

6.1.3. Class IIb Medical Device Conformity Assessment Routes

In the case of devices falling within Class IIb, (other than devices which are custom-made or intended for clinical investigations), the manufacturer shall, either:

1. Follow the procedure relating to the EC declaration of conformity set out in Annex II (full quality assurance); in this case, point 4 of Annex II is not applicable;
2. Follow the procedure relating to the EC type-examination set out in Annex III, coupled with:
   i. The procedure relating to the EC verification set out in Annex IV;
   ii. The procedure relating to the EC declaration of conformity set out in Annex V (production quality assurance); or
   iii. The procedure relating to the EC declaration of conformity set out in Annex VI (product quality assurance).

There are two routes:

1. A Notified Body must carry out either an Annex II audit of the full quality assurance system (ISO 13485).
2. A type-examination (Annex III) plus one of the three options given here:
   i. Examination and testing of each product or homogenous batch of products (Annex IV);
   ii. Audit of the production quality assurance system (Annex V :) ISO 13485 (excluding Design) or

iii. Audit of final inspection and testing (Annex VI :) ISO 13485 (excluding design and manufacture).

Once the manufacturer has received certification from the Notified Body, they may CE mark the product and place them on the market (Figure 6.3).

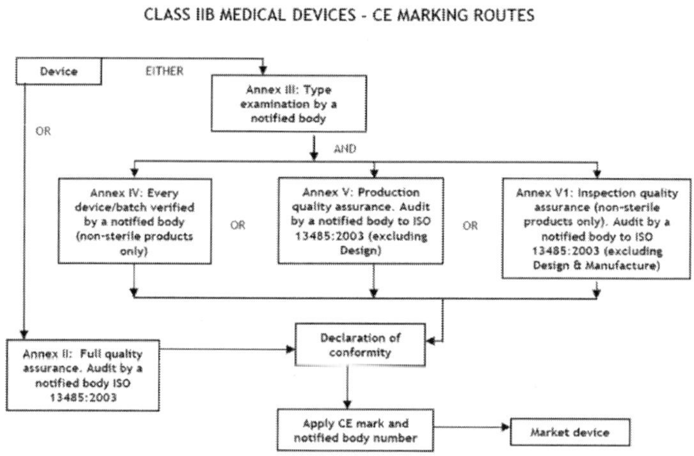

Figure 6.3. Class IIb medical devices CE marking routes.

Simplified Steps for Class IIb Medical Device Compliance	
1.	*Classification: ensure the device is a Class IIb medical device;*
2.	*Choose conformity assessment route;*
3.	*Compile the technical file;*
4.	*Obtain certification from a notified body;*
5.	*Draw up the declaration of conformity;*
6.	*Appoint an authorized representative;*
7.	*Hold the tech files for inspection by the competent authority;*
8.	*Affix CE marking and market the products;*
9.	*Maintain vigilance monitoring and post market surveillance.*

6.1.4. Class III Medical Device Conformity Assessment Routes

This route to conformity is similar to those for Class IIb devices but additionally requires the manufacturer to submit the design dossier to the NB for approval under audit of the full quality assurance system (Annex II) and do not allow the Annex VI option. Active implantable medical devices (AIMD) and *in vitro* diagnostic medical devices (IVD) are also subject to conformity procedures. In the case of devices, falling within Class III (other than devices which are custom-made or intended for clinical investigations) the manufacturer shall either:

1. Follow the procedure relating to the EC declaration of conformity set out in Annex II (full quality assurance);
2. Follow the procedure relating to the EC type-examination set out in Annex III, coupled with:
   i. The procedure relating to the EC verification set out in Annex IV;
   ii. The procedure relating to the EC declaration of conformity set out in Annex V (production quality assurance).

Class III controls are similar to those for Class IIb devices but additionally require the manufacturer to submit the design dossier to the Notified Body for approval under Annex II and do not allow the Annex III/Annex VI option.

There are two routes:

1. A Notified Body must carry out either an Annex II audit of the full quality assurance system (ISO 13485), plus that the manufacturer must submit the design dossier to the Notified Body for approval under Annex II, or
2. A type-examination (Annex III) plus one of the two options given here:
   i. Examination and testing of each product or homogenous batch of products (Annex IV); or
   ii. Audit of the production quality assurance system (Annex V :) ISO 13485 (excluding design)

Once the manufacturer has received certification from the Notified Body they may CE mark the product and place them on the market (Figure 6.4).

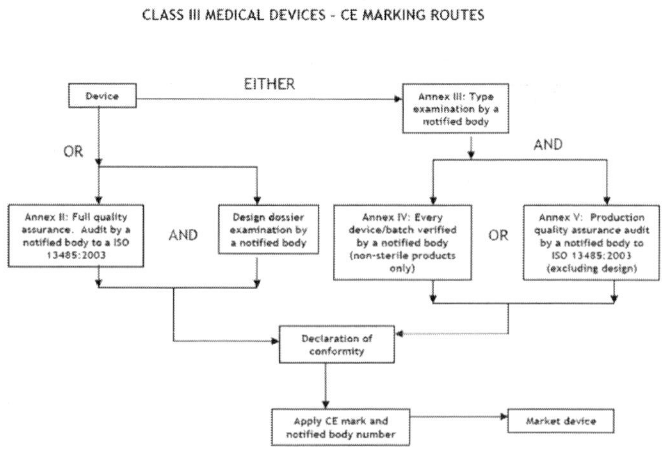

Figure 6.4. Class III medical devices CE marking routes.

Simplified Steps for Class IIb Medical Device Compliance	
1.	*Classification: ensure the device is a Class III medical device;*
2.	*Choose conformity assessment route;*
3.	*Compile the technical file;*
4.	*Obtain certification from a notified body;*
5.	*Draw up the declaration of conformity;*
6.	*Appoint an authorized representative;*
7.	*Hold the tech files for inspection by the competent authority;*
8.	*Affix CE marking and market the products;*
9.	*Maintain vigilance monitoring and post market surveillanc*e.

6.1.5. IVD-In Vitro Diagnostic Medical Device Conformity Assessment Routes

1. For all devices referred to in List an in Annex II other than those intended for performance evaluation, the manufacturer shall, in order to affix the CE marking either:
   i. Follow the procedure relating to the EC declaration of conformity set out in Annex IV (full quality assurance), or
   ii. Follow the procedure relating to EC type-examination set out in Annex V coupled with the procedure relating to the EC declaration of conformity set out in Annex VII (production quality assurance).
2. For all devices referred to in List B in Annex II other than those intended for performance evaluation, the manufacturer shall for

the purposes of affixing the CE marking, follow either:

i. The procedure relating to the EC declaration of conformity set out in Annex IV (full quality assurance); or

ii. The procedure relating to EC type-examination set out in Annex V couplet with:

a. The procedure relating to EC verification set out in Annex VI, or

b. The procedure relating to the EC declaration of conformity set out in Annex VII (production quality assurance).

3. Other than those covered by Annex II.

i. Self-Testing: For all devices for self-testing other than those covered by Annex II and devices for performance evaluation, the manufacturer shall, prior to the drawing up the EC declaration of conformity in Annex III, fulfill the supplementary requirements set out in Annex III, point 6. Instead of applying this procedure, the manufacturer may follow the procedure referred to for products in List A or List B in Annex II.

ii. Other/General: For all devices other than those covered by Annex II and devices for performance evaluation, the manufacturer shall, in order to affix the CE marking, follow the procedure referred to in Annex III and draw up the EC declaration of conformity required before placing the devices on the market.

4. In the case of devices for performance evaluation, the manufacturer shall follow the procedure referred to in Annex VIII and draw up the statement set out in that annex before such devices are made available.

This provision does not affect national regulations relating to the ethical aspects of carrying out performance evaluation studies using tissues or substances of human origin.

6.2. UNDERSTANDING THE CHANGES IN CONFORMITY ASSESSMENTS UNDER THE MDR

Under the MDR, conformity assessment procedures remain essentially the same. Manufacturers still need to conduct a conformity assessment before placing their devices on the market. Similarly, for higher risk devices, the involvement of a notified body is also required. There is also an added requirement for the notified bodies (NBs) to consult with health authorities and EU expert panels and this is expected to prolong quite remarkably the conformity assessment route for class III devices. Apart from Annex numbers being changed, the approach for conducting the conformity assessment routes remains essentially similar to that of the MDD. The number of General Safety and Performance Requirements (GSPRs) have also increased under the MDR in comparison to the ER under the MDD but this will be discussed further later on in this volume.

Device classification will determine the conformity assessment route chosen under the MDR. Classification of a device is regulated by Annex VIII of the MDR and the device itself can be grouped into one of four classes: I, IIa, IIb, and III; taking into account the intended purpose and related risks. Depending on the risk class chosen, the manufacturer may choose the appropriate conformity assessment route to demonstrate compliance with the regulation. Three levels of control are presented and this is dependent on the perceived risk associated with the device:

1. "Self-certification" where no external party is involved in the conformity assessment;
2. Certification by a NB;
3. The involvement of a CA or expert panels through a commission or medical device coordination group (MDCG).

In the MDR, article 52 outlines the conformity assessment procedures as well as the respective annexes which have essentially been renumbered and renamed between Annex IX to XI. Some of the major changes to conformity assessment routes between the MDD and MDR are listed below:

- A new class for reusable surgical instruments (class Ir) has been introduced to control the safety around reusable characteristics of surgical instruments. This class also requires a notified body to assess the quality management system (QMS).
- The MDR introduces new classification rules which may essentially lead to a higher classification of devices than that defined by the MDD. This means there will be an increase in the amount of clinical data required as well as more rigorous post market surveillance.
- The full quality assurance route is replaced by the conformity assessment based on quality system assurance and assessment of

the technical documentation in Annex IX.

- Conformity assessment procedures will generally be more demanding on a manufacturer and also more likely to take longer especially in those cases where a notified body has to consult with health authorities and EU expert panels for the assessment of high-risk devices.

6.3. UNDERSTANDING THE CONFORMITY ASSESSMENT ROUTES IN THE MDR

1. **Annex IX (QMS and Technical Documentation):** Annex IX is used when a full QMS is implemented by the manufacturer. With this conformity assessment route, a review of technical documentation is needed with or without the issuance of a certificate. Depending on the device classification, full or partial compliance with the Annex IX is required.

Annex IX comprises of three parts: 1) Assessment of QMS via NB audit with issuance of CE Marking certificate 2) Assessment of Technical File/Design Dossier via NB audit with issuance of CE Marking certificate 3) General provisions for record retention and availability of documentation to Competent Authority (CA).

2. **Annex X (Type-Examination):** Annex X is used when a manufacturer wants to certify a device based on a representative sample. The NB examines and/or tests the representative sample and associated technical documentation to determine if the device meets MDR requirements with focus on the GSPR.
3. **Annex XI (Product Conformity Verification):** Annex XI is generally used in association with Annex X or in combination with technical documentation annexes (Annexes II and III) for low-risk devices. This annex is composed of two parts: 1) Production Quality Assurance via NB audit with issuance of CE Marking certificate for ability to produce and test a device 2) Product Verification via NB audit with issuance of CE Marking certificate supporting the conformity of a specific batch of devices.

A technical file (or technical documentation) is required to show compliance with Annex II (technical documentation) and Annex III (technical documentation on Post-Market Surveillance-PMS) but this will depend on the device and route of compliance (Table 6.2).

Table 6.2. Comparison of Conformity Assessment Routes between the MDD and MDR

MDD	MDR	Focus of Annex
Annex II excluding section 4. Full quality assurance	Annex IX chapters I and III Quality management system	QMS based Design, manufacture, final inspection
Annex II section 4. Design examination	Annex IX Chapter II Technical documentation	Product based. Type-testing and documentation review
Annex IV verification	Annex IX-Part B Product verification	Product-based Individual devices tested
Annex V Production quality assurance	Annex XI-Part A Production quality assurance	Q<S based, manufacture, final inspection
Annex VI product quality assurance	No equivalent	QMS based, Final inspection
Annex VII Declaration of conformity	Article 19 + annex II, III	For class I devices

6.3.1. Class I Medical Device Conformity Assessment Route Under MDR

Under the MDR, the only conformity assessment route for a self-certified Class I medical device is to maintain technical documentation in compliance with Annex II and III. For Class I devices supplied sterile (Is), with a measuring function (Im), or those that are reusable surgical instruments (Ir), a limited QMS must be in place to control the production (as per Annex XI Part A of the MDR) or to control the special characteristic features (e.g., sterility, measuring, reusability) (Annex IX-Chapter I). For class Is, Im, and Ir devices, an NB will be involved to control how the QMS manages those specific features in relation to the conformity assessment procedure selected (Figure 6.5).

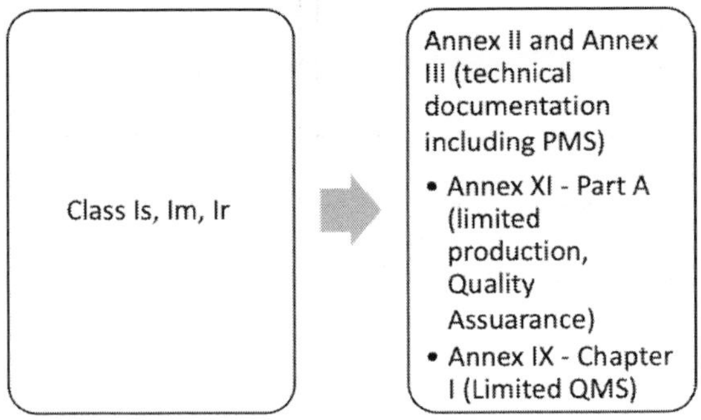

Figure 6.5. Conformity assessment procedures for class I devices under MDR 2017/745.

6.3.2. Class IIa Medical Device Conformity Assessment Route under MDR

Under the MDR, the conformity assessment procedure for a Class IIa device through the review of full QMS (Annex IX) is identical to the procedure for a Class IIb non-active and non-implantable device. Alternatively, a manufacturer may build technical documentation aligned with Annex II and III and select a route of conformity assessment based on the production control (Annex XI) (Figure 6.6).

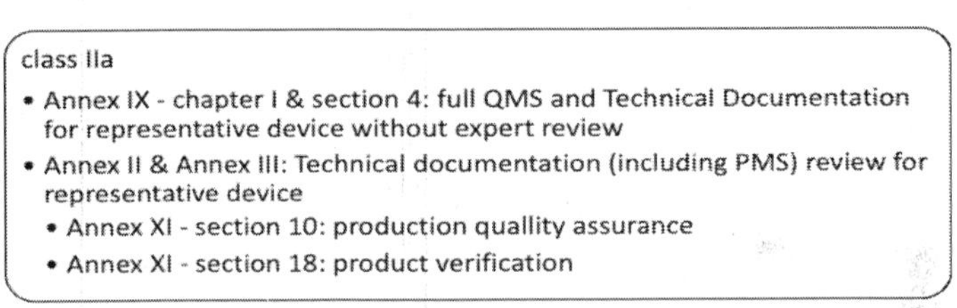

Figure 6.6. Conformity assessment procedures for class IIa devices under MDR 2017/745.

6.3.3. Class IIb Medical Device Conformity Assessment Route under MDR

The level of control for Class IIb devices is similar to Class III devices, though there is no separate CE certificate issued for the technical documentation assessment. In addition, involvement from an expert panel is not required in the review process unless the Class IIb device is active and intended to remove or administer a medicinal substance. The technical documentation review is performed on a representative sample of device type with the exception of Class IIb implantable devices, for which 100% review is required (Figure 6.7).

class IIb active with drug administration or removal

- Annex IX - chapter I & section 4: full QMS and Technical documentation for every device with expert review
- Annex X: assessment of type-examination
  - Annex XI - part A: production quality assurance
  - Annex XI - part B: product verification

class IIb

- Annex IX - chapter I & section 4: for class IIb implantable devices*: full QMS and technical documentation for every device without expert review/for class IIb non-implantable devices*: full QMS and technical documentation for representative device without expert review
- Annex X: assessment of type examination
  - Annex XI - part A: product quality assurance
  - Annex XI - part B: product verification

*except sutures, staples, dental fillings, dental braces, tooth crowns, screws, wedges, plates, wires, pins, clips, and connectors.

Figure 6.7. Conformity assessment procedures for class IIb devices under MDR 2017/745.

6.3.4. Class III Medical Device Conformity Assessment Route Under MDR

Conformity assessment for Class III medical devices is covered by Annex IX and includes a full Quality Assurance audit and full Technical Documentation review. A certificate is issued following the NB QMS audit and a second certificate is issued for every device after review of the technical documentation. An expert panel is also involved in the evaluation for Class III implantable devices.

While standards are voluntary, one way of presuming conformity to the GSPR and meeting the provisions of full quality assurance is to possess harmonized EN ISO 13485 standard certifications, which pertains to the state of the art and expected requirements for a company's QMS in the EU. It should be noted that manufacturers of Class III medical devices also have the option to pursue Annex X, Type-Examination, in combination with Annex XI Part A or Part B and, therefore, with a QMS focused on production and controls (Figure 6.8).

class III

- Annex IX: assessment of full QMS and technical documentation for every device with e xpert review
- Annex X: assessment of type examination
  - Annex XI - part A: production quality assurance
  - Annex IX - part B: product verification

Figure 6.8. Conformity assessment procedures for class III devices under MDR 2017/745.

CHAPTER

7

General Safety and Performance Requirements

CONTENTS

While compliance with the ER is imperative for establishing conformity with the MDD and active implantable medical device directive (AIMDD, 90/385/EEC), the same applies to the General Safety and Performance Requirements (GSPR) within the MDR. The main purpose of this chapter is to compare the ER within the MDD to that of the GSPR of the MDR.

The first point to note is that the GSPRs are detailed in Annex I of the MDR and are structured into three chapters, namely:

- **Chapter I:** General requirements.
- **Chapter II:** Requirements regarding design and manufacture.
- **Chapter III:** Requirements regarding the information supplied with the device.

While there are 13 ER in the MDD and 16 in the AIMDD, there are 23 GSPRs in the new MDR. The overall requirements have been expanded, but the scope is the same as that of the previous directives. There are a few notable exceptions. For example, clinical evaluation has moved into the articles section of the regulation and a few new topics have been added such as devices without a medical purpose and requirements for devices used by laypersons. In addition, some parts of the GSPR now have increased emphasis and more detailed requirements so as to allow for more thorough alignment with harmonized standards and industry guidance. Some numbering and organizational changes have taken place, the requirements for risk and labeling have been expanded, and there are new areas of emphasis as per standards and industry guidance, and many "state of the art" requirements from harmonized standards have been incorporated directly into the Regulation's GSPRs.

Let's take a closer look at some of the detail of the GSPR:

7.1. GSPR 1: PERFORMANCE AND SAFETY

GSPR 1 generally corresponds to MDD ER 1 with the majority of the text being the same. This requirement states that the devices shall be 'designed and manufactured in such a way' that safety of patients and users shall not be compromised. As with ER 1 in the directives, this is under the normal conditions of use. The concept of 'performance' is also brought in from the MDD ER 3. Design and construction of the device should conform to safety principles, taking into account the 'generally acknowledged state of the art' as required in to MDD ER 2. Risks relating to ergonomic features and consideration of the use environment which was previously present in MDD ER 1 have now moved to GSPR 5.

7.2. GSPR 2: REDUCTION OF RISKS

The GSPR 2–5 all relate to risk management. GSPR 2 is a new statement addressing the requirement to reduce risks as far as possible without adversely affecting the risk-benefit ratio.

7.3. GSPR 3: RISK MANAGEMENT SYSTEM

GSPR 3 is a new requirement which more explicitly defines the requirements of risk management as aligned with EN ISO 14971.

7.4. GSPR 4: RISK CONTROL MEASURES AND RESIDUAL RISKS

GSPR 4 generally corresponds to ER 2 in the existing MDD and covers the requirements that both, the overall residual risks and the residual risk associated with each hazard is evaluated and judged to be acceptable, with respect to the benefits. While this is already an expectation of the harmonized standard, this will now be explicitly included in the regulation. Furthermore, there is also a requirement that the manufacturer 'shall inform users of any residual risks.'

7.5. GSPR 5: RISKS RELATED TO USE

GSPR 5 corresponds to the latter part of ER 1 in the MDD and addresses the reduction of risks relating to use error. Usability or human factors are addressed such that the risk of use error shall be reduced as far as possible.

7.6. GSPR 6: DEVICE LIFETIME

GSPR 6 defines requirements for the characteristics and device performance over the lifetime of the device as indicated. The device characteristics and performance shall not be adversely impacted to such a degree that health or safety of a patient or user would be compromised, when under the stresses of 'normal conditions of use.' Device performance must not be adversely impacted when the device 'has been properly maintained in accordance with the manufacturer's instructions.' Reasonably foreseeable maintenance or storage errors may be considered as part of use risk evaluation, but in some cases, impacts to performance cannot be prevented when maintenance and storage are not per the manufacturer's instructions.

7.7. GSPR 7: PACKAGING, TRANSPORT, STORAGE

This requirement outlines the basics for device packaging, transportation, and storage. This requirement can be linked to ER 5 in the MDD. The regulation provides examples of potential temperature and humidity fluctuations which are commonly considered by manufacturers during transportation simulations and storage where necessary.

7.8. GSPR 8: RISK-BENEFIT RATIO

Here an updated definition of the 'risk-benefit ratio' is provided. It now includes 'all known and foreseeable risks,' and states that the risks 'shall be minimized.'

7.9. GSPR 9: DEVICES WITHOUT A MEDICAL PURPOSE

Devices without a medical purpose are excluded from the scope of the MDD but are included in the MDR. GSPR 9 clarifies how to apply the 'risk-benefit' and 'performance' requirements for those devices without a medical purpose or medical benefit.

7.10. GSPR 10: CHEMICAL, PHYSICAL, AND BIOLOGICAL PROPERTIES

The requirements of GSPR 10 are broadly related to biological safety with more focus on the areas of specific substances of concern.

7.11. GSPR 11: INFECTION AND MICROBIAL CONTAMINATION

The requirements of GSPR 11 are generally similar to that of MDD ER 8 and relate to infection and microbial contamination. This includes risk of infection, design for reuse, devices with a specific microbial state, devices delivered sterile, validation for sterile devices, environmental controls, packaging for non-sterile devices, and labeling for sterile state.

7.12. GSPR 12: DEVICES INCORPORATING A MEDICINAL PRODUCT, SUBSTANCES ABSORBED OR LOCALLY DISPERSED

GSPR 12 deals with the requirements for medicinal products or substances which are absorbed or locally dispersed in the body. Medicinal substances were previously addressed in MDD ER 7.4. It should be noted that the scope of GSPR 12 has significantly changed under MDR with a broader scope with regards to the definition of what is included along with the medicinal considerations which now includes 'substances or combinations of substances that are absorbed by or locally dispersed in the human body.'

7.13. GSPR 13: DEVICES INCORPORATING MATERIALS OF BIOLOGICAL ORIGIN

Materials of biological origin within the MDR have expanded a great deal, from what was originally included in the MDD. These requirements generally correspond to MDD ER7.4 and 8.2 and address tissues, cells or derivatives of human origin.

7.14. GSPR 14: CONSTRUCTION OF DEVICES AND INTERACTION WITH THEIR ENVIRONMENT

GSPR 14 corresponds to ER 9 in the MDD as well as ER 10.2 for ergonomic principles and deals with the construction of devices and interaction with their environment. Specifically, it focuses on use in combination, risks of interaction with the environment, risks of fire or explosion, design for adjustment, calibration, and maintenance; design for compatibility, measurement, monitoring or display scales and design and manufacture for safe disposal.

7.15. GSPR 15: DEVICES WITH A DIAGNOSTIC OR MEASURING FUNCTION

GSPR 15 of the MDR (in comparison to the MDD) deals with 'diagnostic devices' which has been added to the scope of the requirement. Manufacturers should consider the 'precision' of the device and measurements from measuring devices are still required to be expressed in legal units as per Directive 80/181/EEC.

7.16. GSPR 16: PROTECTION AGAINST RADIATION

These requirements in GSPR 16 generally correspond to parts of ER 11 in the MDD and covers protection against radiation. This includes general requirements, requirements for intended and unintended radiation as well as ionizing radiation.

7.17. GSPR 17: ELECTRONIC PROGRAMMABLE SYSTEMS AND SOFTWARE

GSPR 17 defines the requirements for electronic programmable systems and software considered to be a medical device. The specific wording of the Regulation now includes 'software that are devices in themselves' more explicitly in the scope of these requirements. Information security is also included.

7.18. GSPR 18: ACTIVE DEVICES AND DEVICES CONNECTED TO THEM

GSPR 18 defines particular requirements for active devices and connected devices with a large portion of the requirement corresponding to subparts of ER 12 in the MDD.

7.19. GSPR 19: PARTICULAR REQUIREMENTS FOR ACTIVE IMPLANTABLE DEVICES

GSPR 19 defines the requirements for active implantable devices. The requirements can be generally linked to ERs in the current AIMDD however, as the MDD and AIMDD requirements have now been merged into the single MDR, many of the requirements for active implantable devices have moved and are listed here. This includes particular risks to be reduced for active implantable devices, device compatibility, and reliability of energy, identification of devices and components and identification codes.

7.20. GSPR 20: PROTECTION AGAINST MECHANICAL AND THERMAL RISKS

The requirements in GSPR 20 can be directly correlated to ER 12.7 in the MDD. This includes errors that could be made in fitting or refitting parts. This is usually covered as manufacturers align to the current standards and usability requirements.

7.21. GSPR 21: PROTECTION AGAINST THE RISKS POSED TO THE PATIENT OR USER BY DEVICES SUPPLYING ENERGY OR SUBSTANCES

GSPR 21 corresponds to ERs 12.8.1, 12.8.2, and 12.9 in the MDD and deals with protection against the risks posed to the patient or user by devices supplying energy or substances. There is some minor rewording of the requirement however, the intent is unchanged.

7.22. GSPR 22: PROTECTION AGAINST THE RISKS POSED BY MEDICAL DEVICES INTENDED BY THE MANUFACTURER FOR USE BY LAY PERSONS

Specific requirements for devices used by laypersons are new in the MDR. GSPR 22 outlines the requirements and considerations for the device's instructions, variation in user technique, risks of error and injury, and verification of device function by the user.

7.23. GSPR 23: LABEL AND INSTRUCTIONS FOR USE

GSPR 23 deals with product labels and IFU in which the text and requirements for labels and IFUs are greatly expanded as compared with both Directives. This includes the general requirements for information supplied by the manufacturer, label requirements as well as sterile package label requirements and requirements pertaining to the

CHAPTER

8

CONCLUSION

In recent times, the need for improving regulatory controls and finding ways to improve the quality and access to safe medical devices has increased. The establishment of an effective regulatory system is dependent on numerous factors including long term government commitment and an overall political and societal willingness to cooperate. As you would have now seen, the new medical device regulation (MDR 2017/745) in not entirely different from the medical device directive (MDD) as many of the related standards have been continuously updated in the last 25 years since the MDD existed. For this reason, much of the changes in the new regulation are essentially a more stringent or stricter form of previous requirements. Changes brought about by the MDR in Europe require manufacturers to take a proactive, multilevel, and multidisciplinary approach when it comes to implementation. Manufacturers will need to start by reviewing their product portfolios in the context of the MDR and assessing their overall readiness while prioritizing the time-consuming areas of the transition such as designating a suitable person for regulatory control, aligning to the additional requirements for clinical evaluations, and implementing a more detailed post market surveillance system. Impacts will be felt on all levels of the business, from supply chain to marketing, and for this reason ensuring a thorough understanding of the MDR as well as being able to properly identify its full impact on the business is critical to a successful and effective transition.

Appendices

Appendix 1: Template for an Essential Requirements Checklist (under MDD)

No.	Essential Requirements	A or N/A	Relevant Standards	Evidence Showing Compliance	Comment
1.	**General Requirements** The devices must be designed and manufactured in such a way that, when used under the conditions and for the purposes intended, they will not compromise the clinical condition or the safety of patients, or the safety and health of users or, where applicable, other persons, provided that any risks which may be associated with their intended use constitute acceptable risks when weighed against the benefits to the patient and are compatible with a high level of protection of health and safety. This shall include: • Reducing as far as possible, the risk of use error due to the ergonomic features of the device and the environment in which the device is intended to be used (design for patient safety), and • Consideration of the technical knowledge, experience, education, and training, and where applicable the medical and physical conditions of intended users (design for lay, professional, disabled or other users).				

2.	The solutions adopted by the manufacturer for the design and construction of the devices must conform to safety principles, taking account of the generally acknowledged state of the art. In selecting the most appropriate solutions, the manufacturer must apply the following principles in the following order: • Eliminate or reduce risks as far as possible (Inherently safe design and construction). • Where appropriate take adequate protection measures including alarms if necessary, in relation to risks that cannot be eliminated. • Inform users of the residual risks due to any shortcomings of the protection methods adopted.				
3.	The devices must achieve the performance intended by the manufacturer and be designed, manufactured, and packaged In such a way that they are suitable for one or more of the functions referred to in Article 1(2) (a) as specified by the manufacturer				
4.	The characteristics and performances referred to In sections 1, 2 and 3 must not be adversely affected to such a degree that the clinical condition and safety of the patients and, where applicable, of other persons are compromised during the lifetime of the device as Indicated by the manufacturer, when the device Is subjected to the stresses which can occur during normal conditions of use.				
5.	The devices must be designed, manufactured, and packed in such a way that their characteristics and performances during their intended use will not be adversely affected during transport and storage taking account of the instructions and Information provided by the manufacturer				
6.	Any undesirable side effects must constitute an acceptable risk when weighed against the performances intended.				
	a. Demonstration of conformity with the essential requirements must include a clinical evaluation in accordance with Annex X.				
7.	Requirements regarding design and construction chemical. Physical and biological properties				
7.1	The devices must be designed and manufactured in such a way as to guarantee the characteristics and performances referred to in Section I on the "General requirements." Particular attention must be paid to: • The choice of materials used, particularly as regards toxicity and, where appropriate flammability; • The compatibility between the materials used and biological tissues, cells, and body fluids, taking account of the intended purpose of the device; • Where appropriate, the results of biophysical or modeling research whose validity has been demonstrated beforehand.				

7.2	The devices must be designed, manufactured, and packed In such a way as to minimize the risk posed by contaminants and residues to the persons involved in the transport, storage, and use of the devices and to the patients, taking account of the intended purpose of the product. Particular attention must be paid to the tissues exposed and the duration and frequency of the exposure.				
7.3	The devices must be designed and manufactured in such a way that they can be used safely with materials, substances, and gases with which they enter into contact during normal use or during routine procedures; if the devices are Intended to administer medicinal products they must be designed and manufactured in such a way as to be compatible with the medicinal products concerned according to the provisions and restrictions governing those products and that their performance is maintained in accordance with the Intended use.				
7.4	Where a device incorporates, as an integral part, a substance which, if used separately, may be considered to be a medicinal product as defined in Article 1 of Directive 2001/83/EC and which is liable to act upon the body with action ancillary to that of the device, the quality, safety, and usefulness of the substance must be verified by analogy with the methods specified in Annex I of Directive 2001/83/EC. For the substances referred to in the first paragraph, the notified body shall, having verified the usefulness of the substance as part of the medical device and taking account of the intended purpose of the device, seek a scientific opinion from one of the competent authorities designated by the Member States or the European Medicines Agency (EMEA) acting particularly through its committee in accordance with Regulation (EC) No. 726/2004 (*) on the quality and safety of the substance including the clinical benefit/risk profile of the incorporation of the substance into the device. When issuing its opinion, the competent authority or the EMEA shall take into account the manufacturing process and the data related to the usefulness of the incorporation of the substance into the device as determined by the notified body.				
7.4	Where a device incorporates, as an integral part, a human blood derivative, the notified body shall, having verified the usefulness of the substance as part of the medical device and taking into account the purpose of the device, seek a scientific opinion from the EMEA, acting particularly through its committee, on the quality and safety of the substance including the clinical benefit/risk profile of the incorporation of the human blood derivative into the device. When issuing its opinion, the EMEA shall take into account the manufacturing process and the data related to the usefulness of incorporation of the substance into the device as determined by the notified body.				

	Where changes are made to an ancillary substance incorporated in a device, in particular related to its manufacturing process, the notified body shall be informed of the changes and shall consult the relevant medicines competent authority (i.e., the one involved in the initial consultation), in order to confirm that the quality and safety of the ancillary substance are maintained. The competent authority shall take into account the data related to the usefulness of incorporation of the substances into the device as determined by the notified body, in order to ensure that the changes have no negative impact on the established benefit/risk profile of the addition of the substance in the medical devices.				
7.4	(*) Regulation (EC) No. 725/2994 of the European Parliament and of the Council of 31 March 2004 laying down Community procedures for the authorization and supervision of medicinal products for human and veterinary use and establishing a European Medicines Agency (OJ L 136, 30.4.2004, p.1). Regulation as last amended by Regulation (EC) No. 1901/2006.				
7.4	When the relevant medicines competent authority (i.e., the one involved in the initial consultation) has obtained information on the ancillary substance, which could have an impact on the established benefit/risk profile of the addition of the substance in the medical device, it shall provide the notified body with advice, whether this information has an impact on the established benefit/risk profile of the addition of the substance in the medical device or not. The notified body shall take the updated scientific opinion into account in reconsidering its assessment of the conformity assessment procedure.				
7.5	The devices must be designed and manufactured in such a way as to reduce to a minimum the risks posed by substances leaking from the device. Special attention shall be given to substances which are carcinogenic, mutagenic, or toxic to reproduction, in accordance with Annex I to Council Directive 67/548/EEC of 27 June 1967 on the approximation of laws, regulations, and administrative provisions relating to the classification, packaging, and labeling of dangerous substances. If parts of a device (or a device itself) intended to administer and/or remove medicines, body liquids or other substances to or from the body, or devices intended for transport or storage of such body fluids or substances, contain phthalates which are classified as carcinogenic, mutagenic or toxic to reproduction, of category 1 or 2, in accordance with Annex I of Directive 67/548/EEC, these devices must be labeled on the device itself and /or on the packaging for each unit or, where appropriate, on the sales packaging as a device containing phthalates.				

	If the intended use of such devices includes treatment of children or treatment of pregnant or nursing women, the manufacturer must provide a specific justification for the use of these substances with regard to compliance with the essential requirements, in particular of this paragraph, within the technical documentation and, within the instructions for use, information on residual risks for these patient groups and, if applicable, on appropriate precautionary measures. (*) OJ 196, 16.8.1967, p 1. Directive as last amended by Directive 2006/121/EC of the European Parliament and of the Council (OJ) L 396, 30.12.2006, p. 850				
7.6	The devices must be designed and manufactured In such a way as to reduce as much as possible, risks posed by the unintentional ingress of substances into the device taking into account the device and the nature of the environment in which it Is intended to be used.				
8.	Infection and microbial contamination				
8.1	The devices and their manufacturing processes must be designed in such a way as to eliminate or reduce as far as is possible the risk of infection to the patient, user, and third parties. The design must allow easy handling and, where necessary, minimize contamination of the device by the patient or vice versa during use.				
8.2	Tissues of animal origin must originate from animals that have been subjected to veterinary controls and surveillance adapted to the intended use of the tissues. Notified bodies (NBs) shall retain information on the geographical origin of the animals. Processing, preservation, testing, and handling of tissues, cells, and substances of animal origin must be carried out so as to provide optimal security. In particular, safety with regard to viruses and other transmissible agents must be addressed by implementation of validated methods of elimination or viral inactivation in the course of the manufacturing process.				
8.3	Devices delivered in a sterile state must be designed, manufactured, and packed in a non-reusable pack and/or according to appropriate procedures to ensure they are sterile when placed on the market and remain sterile, under the storage and transport conditions laid down, until the protective packaging is damaged or opened.				
8.4	Devices delivered in a sterile state must have been manufactured and sterilized by an appropriate, validated method.				
8.5	Devices intended to be sterilized must be manufactured in appropriately controlled (e.g., environmental) conditions.	N/A			
8.6	Packaging systems for non-sterile devices must keep the product without deterioration in the level of cleanliness stipulated and, if the devices are to be sterilized prior to use, minimize the risk of microbial contamination. The packaging system must be suitable taking account of the method of sterilization indicated by the manufacturer.				
8.7	The packaging and/or label of the device must distinguish between identical or similar products sold in both sterile and non-sterile condition.				

9.	Construction and environmental Properties				
9.1	If the device is intended for use in combination with other devices or equipment, the 'whole combination, including the connection system must be safe and must not impair the specified performance of the devices. Any restrictions on use must be indicated on the label or in the Instruction for Use.				
9.2	Devices must be designed and manufactured in such a way as to remove or minimize as far as possible: the risk of injury in connection with their physical features [1] including the volume pressure ratio, dimensional, and 'where appropriate the ergonomic features. - Risks connected with reasonably foreseeable environmental conditions, such as magnetic fields external electrical influences, electrostatic discharge, pressure, temperature or variations in pressure and acceleration, the risks of reciprocal interference with other devices normally used in the investigations or for the treatment given, risks arising where maintenance or calibration are not possible (as with implants) from aging of the materials used or loss of accuracy of any measuring or control mechanism.				
9.3	Devices must be designed and manufactured In such a way as to minimize the risks of fire or explosion during normal use and in single fault condition. Particular attention must be paid to devices 'whose intended use includes exposure to flammable substance 'which could cause combustion.				
10.	Devices with a measuring function				
10.1	Devices with a measuring function must be designed and manufactured in such a way as to provide sufficient accuracy and stability 'within appropriate limits of accuracy and taking account of the Intended purpose of the device. The limits of accuracy must be indicated by the manufacturer.				
10.2	The measurement, monitoring, and display scale must be designed In line 'with ergonomic principles, taking account of the intended purpose of the device.				
10.3	The measurements made by devices 'with a measuring function must be expressed in legal units conforming to the provisions of Council Directive *8OII8IIEEC,* as last amended by Directive 89/17/EC.				
11.	Protection against radiation				
11.1	General: Devices shall be designed and manufacturer such that exposure of patients, users, and other persons to radiation shall be reduced as far as possible compatible 'with the intended purpose, whilst not restricting the application of appropriate specified levels for therapeutic and diagnostic purposes.				
11.2	Intended radiation				
11.2.1	Where devices are designed to emit hazardous levels of radiation necessary for a specific medical purpose the benefit of 'which Is considered to outweigh the risks inherent in the emission, it must be possible for the user to control the emissions. Such devices shall be designed and manufactured to ensure reproducibility and tolerance of relevant variable parameters				

11.2.2	Where devices are intended to emit potentially hazardous, visible, and/or invisible radiation, they must be fitted, 'where practicable, 'with visual displays and/or audible warnings of such emissions.				
11.3	Unintended radiation: Devices shall be designed and manufactured in such a way that exposure of patients, users, and other persons to the emission of unintended, stray, or scattered radiation must be reduced as far as possible.				
11.4	Instructions: The operating instructions for devices emitting radiation must give detailed information as to the nature of the emitted radiation, means of protecting the patient and the user and on ways of avoiding misuse and of eliminating the risks inherent in installation				
11.5	Ionizing radiation				
11.5.1	Devices Intended to emit ionizing radiation must be designed and manufactured in such a way as to ensure that, 'where practicable, the quantity, geometry, and quality of radiation emitted can be varied and controlled taking account of the Intended uses.				
11.5.2	Devices emitting ionizing radiation intended for diagnostic radiology shall be designed and manufactured In such a way, as to achieve appropriate Image and/or output quality for the intended medical purpose whilst minimizing radiation exposure of the patient and user				
11.5.3	Devices emitting ionizing radiation, Intended for therapeutic radiology shall be designed and manufactured In such a way as to enable reliable monitoring and control of the delivered dose, the beam type and energy and 'where appropriate the quality of the radiation.				
12.	Requirements for medical devices connected to or equipped with an energy source				
12.1	Devices incorporating electronic programmable systems must be designed to ensure the repeatability. Reliability and performance of these systems according to their intended use. In the event of a single fault condition (in the system), appropriate means should be adopted to eliminate or reduce as far as possible consequent risks.				
	a. For devices which incorporate software or which are medical software in themselves, the software must be validated according to the state of the art taking into account the principles of development lifecycle, risk management, validation, and verification.				
12.2	Devices 'where the safety of the patients depends on an internal power supply must be equipped 'with a means of determining the state of the power supply.				
12.3	Devices where the safety of the patient depends on an external power supply must Include an alarm system to signal any power failure.				
12.4	Devices Intended to monitor one or more clinical parameters of a patient must be equipped 'with appropriate alarm systems to alert the user of situations 'which could lead to death or severe deterioration of the patient's state of health.				
12.5	Devices must be designed and manufactured in such a way as to minimize the risks of creating electromagnetic fields 'which could impair the operation of other devices or equipment In the usual environment				

12.6	Protection against electrical risks. Devices must be designed and manufactured in such a way as to avoid, as far as possible, the risk of accidental electric shocks during normal use and In single fault condition, provided that the devices are Installed correctly.				
12.7	Protection against mechanical and thermal risk				
12.7.1	The devices must be designed and manufactured in such a way as to protect the patient and user against mechanical risks connected 'with, for example, resistance, stability, and moving parts.				
12.7.2	The devices must be designed and manufactured in such a way as to reduce to the lowest possible level the risks arising from vibration generation by the devices, taking account of technical progress and of the means available for limiting vibrations, particularly at source, unless the vibrations are part of the specified performance.				
12.7.3	The devices must be designed and manufactured In such a way as to reduce to the lowest possible level the risks arising from the noise emitted [1] taking account of technical progress end of the means available to reduce noise [1] particularly at source [9] unless the noise emitted Is part of the specified performance.				
12.7.4	The terminals and connections to the electricity gas or hydraulic and pneumatic energy supplies which the user has to handle must be designed and constructed in such a way as to minimize all possible risks.				
12.7.5	Accessible parts of devices (excluding any parts or areas intended to supply heat or reach given temperatures) and their surroundings must not attain potentially dangerous temperatures under normal use.				
12.8	Protection against the risks posed to the patient by energy supplies or substances				
12.8.1	Devices for supplying the patient with energy or substances must be designed and constructed in such a way that the flow rate can be set and maintained accurately enough to guarantee the safety of the patient and the user.				
12.8.2	Devices must be fitted with the means of preventing and/or indicating any inadequacies in the flow-rate, which could pose a danger. Devices must incorporate suitable means to prevent, as far as possible, the accidental release of dangerous levels of energy from an energy and/ or substance source				
12.9	The function of the controls and indicators must be clearly specified on the devices. Where a device beans instruction required for its operation or Indicates operating or adjustment parameters by means of a visual system, such information must be understandable to the user and, as appropriate, the patient				
13.	Information supplied by the manufacturer				

13.1	Each device must be accompanied by the information needed to use it safely and properly, taking account of the training and knowledge of the potential user, and to identify the manufacturer. This information comprises the details on the label and the data in the Instructions for use. As far as practicable and appropriate, the information needed to use the device safely must be set out on the device itself and/or on the packaging for each unit or, where appropriate, on the sales packaging. If individual packaging of each unit is not practicable, the information must be set out in the leaflet supplied with one or more devices. Instructions for use must be included in the packaging for every device. By way of exception, no such instruction leaflet is needed for devices in Class I or Class IIa if they can be used completely safely without any such instructions.				
13.2	Where appropriate this information should take the form of symbols. Any symbol or identification color used must conform to the harmonized standards. In areas for which no standards exist, the symbols and colors must be described in the documentation supplied with the device.				
13.3	The label must bear the following particulars:				
	a. The name or trade name and address of the manufacturer. For devices imported into the Community, in view of their distribution in the Community, the label, or the outer packaging, or the instructions for use, shall contain in addition the name and address of the authorized representative where the manufacturer does not have a registered place of business in the Community;				
	b. The details strictly necessary to identify the device and the contents of the packaging especially for the users;				
	c. Where appropriate the word "STERILE";				
	d. Where appropriate, the batch code preceded by the word "LOT" or the serial number.				
	e. Where appropriate an indication of the date by which the device should be used, in safety, expressed as the year and month;				
	f. Where appropriate, an indication that the device is for single use. A manufacturer's indication of single use must be consistent across the Community;				
	g. If the device is custom-made, the words "custom made device"				
	h. If the device is intended for clinical investigations, the words "exclusively for clinical investigations";				
	i. Any special storage and/or handling conditions;				
	j. Any special operating instructions;				
	k. Any warnings and/or precautions to take;				
	l. Year of manufacture of active devices other than those covered by (e). This indication may be included in the batch or serial number;				
	m. Where applicable, method of sterilization.				

	n. In the case of a device within the meaning of Article 1(4a), an indication that the device contains a human blood derivative.				
13.4	If the Intended purpose of the device is not obvious to the user, the manufacturer must clearly state it on the label and in the instruction leaflet.				
13.5	Wherever reasonable and practicable, the devices and detachable components must be identified, where appropriate in terms of batches, to allow all appropriate action to detect any potential risk posed by the devices and detachable components.				
13.6	Where appropriate, the Instructions for Use must contain the following particulars:				
	a. The details referred to In 13.3 with the exception of (d) and (e)				
	b. The performances referred to in section 3 and any undesirable side effects				
	c. If the device must be installed with or connected to other medical devices or equipment in order to operate as required for its intended purpose, sufficient details of its characteristics to identify the correct devices or equipment to use in order to obtain a safe combination;				
	d. All the information needed to verify whether the device is properly installed and can operate correctly and safely, plus details of the nature and frequency of the maintenance and calibration needed to ensure that the devices operate properly and safely at all times;				
	e. Where appropriate, information to avoid certain risks in connection with implantation of the device;				
	f. Information regarding the risks of reciprocal interference posed by the presence of the device during specific investigations or treatment;				
	g. The necessary instructions in the event of damage to the sterile packaging and, where appropriate, details of appropriate methods of re-sterilization;				
	h. If the device is reusable, information on the appropriate processes to allow reuse, including cleaning, disinfection, packaging, and, where appropriate, the method of sterilization of the device to be re-sterilized, and any restriction on the number of reuses. Where devices are supplied with the intention that they be sterilized before use, the instructions for cleaning and sterilization must be such that, if correctly followed, the device will still comply with the requirements in Section 1; If the device bears an indication that, the device is for single use, information on known characteristics and technical factors known to the manufacturer that could pose a risk if the device were to be re-used. If in accordance with Section 13.1 no instructions for use are needed, the information must be made available to the user upon request;				
	i. Details of any further treatment or handling needed before the device can be used (for example, sterilization, final assembly, etc.);				
	j. In the case of devices emitting radiation for medical purposes, details of the nature, type, intensity, and distribution of this radiation.				

	k. The instruction for use must also include details allowing the medical staff to brief the patient on any contra-indications and any precautions to be taken. These details should cover in particular: – precautions to be taken in the event of changes in the performance of the device;				
	l. The instruction for use must also include details allowing the medical staff to brief the patient on any contra-indications and any precautions to be taken. These details should cover in particular: - precautions to be taken as regards exposure, in reasonably foreseeable environmental conditions, to magnetic fields, external electrical influences, electrostatic discharge, pressure or variations' in pressure, acceleration, thermal ignition sources, etc.				
	m. The instruction for use must also include details allowing the medical staff to brief the patient on any contra-indications and any precautions to be taken. These details should cover in particular: – adequate information regarding the medicinal products which the device in question is designed to administer, including any limitations in the choice of substances to be delivered;			*	
	n. The instruction for use must also include details allowing the medical staff to brief the patient on any contra-indications and any precautions to be taken. These details should cover in particular: – Precautions to be taken against any special unusual risks related to the disposal of the device.				
	o. The instruction for use must also include details allowing the medical staff to brief the patient on any contra-indications and any precautions to be taken. These details should cover in particular: – medicinal substances incorporated into the device as an integral part in accordance with section 7.4;				
	p. The instruction for use must also include details allowing the medical staff to brief the patient on any contra-indications and any precautions to be taken. These details should cover in particular: – degree of accuracy claimed for devices with a measuring function;				
	q. Date of issue or the latest revision of the instructions for use.				

Simple easy to use templates and guides to completing an Essential Requirements checklist can be downloaded from https://scientistssanctuary.com/

Appendix 2: Template for a General Safety and Performance Requirements Checklist (under MDR)

No.	Essential requirements	A or N/A	Relevant Standards	Evidence Showing Compliance	Comment
1.	General Requirements: Devices shall achieve the performance intended by their manufacturer and shall be designed and manufactured in such a way that, during normal conditions of use, they are suitable for their intended purpose. They shall be safe and effective and shall not compromise the clinical condition or the safety of patients, or the safety and health of users or, where applicable, other persons, provided that any risks which may be associated with their use constitute acceptable risks when weighed against the benefits to the patient and are compatible with a high level of protection of health and safety, taking into account the generally acknowledged state of the art.				
2.	The requirement in this Annex to reduce risks as far as possible means the reduction of risks as far as possible without adversely affecting the benefit-risk ratio.				
3.	Manufacturers shall establish, implement, document, and maintain a risk management system. Risk management shall be understood as a continuous iterative process throughout the entire lifecycle of a device, requiring regular systematic updating. In carrying out risk management manufacturers shall: a. Establish and document a risk management plan for each device; b. Identify and analyze the known and foreseeable hazards associated with each device; c. Estimate and evaluate the risks associated with, and occurring during, the intended use and during reasonably foreseeable misuse; d. Eliminate or control the risks referred to in point. c. In accordance with the requirements of Section 4; e. Evaluate the impact of information from the production phase and, in particular, from the post-market surveillance system, on hazards and the frequency of occurrence thereof, on estimates of their associated risks, as well as on the overall risk, benefit-risk ratio and risk acceptability; and f. Based on the evaluation of the impact of the information referred to in point (e), if necessary, amend control measures in line with the requirements of Section 4.				

4.	Risk control measures adopted by manufacturers for the design and manufacture of the devices shall conform to safety principles, taking account of the generally acknowledged state of the art. To reduce risks, Manufacturers shall manage risks so that the residual risk associated with each hazard as well as the overall residual risk is judged acceptable. In selecting the most appropriate solutions, manufacturers shall, in the following order of priority: a. Eliminate or reduce risks as far as possible through safe design and manufacture; b. Where appropriate, take adequate protection measures, including alarms if necessary, in relation to risks that cannot be eliminated; and c. Provide information for safety (warnings/ precautions/contra-indications) and, where appropriate, training to users. Manufacturers shall inform users of any residual risks.				
5.	In eliminating or reducing risks related to use error, the manufacturer shall: a. Reduce as far as possible the risks related to the ergonomic features of the device and the environment in which the device is intended to be used (design for patient safety), and b. Give consideration to the technical knowledge, experience, education, training, and use environment, where applicable, and the medical and physical conditions of intended users (design for lay, professional, disabled or other users)				
6.	The characteristics and performance of a device shall not be adversely affected to such a degree that the health or safety of the patient or the user and, where applicable, of other persons are compromised during the lifetime of the device, as indicated by the manufacturer, when the device is subjected to the stresses which can occur during normal conditions of use and has been properly maintained in accordance with the manufacturer's instructions.				
7.	Devices shall be designed, manufactured, and packaged in such a way that their characteristics and performance during their intended use are not adversely affected during transport and storage, for example, through fluctuations of temperature and humidity, taking account of the instructions and information provided by the manufacturer.				
8.	All known and foreseeable risks, and any undesirable side-effects, shall be minimized and be acceptable when weighed against the evaluated benefits to the patient and/or user arising from the achieved performance of the device during normal conditions of use.				
9.	For the devices referred to in Annex XVI, the general safety requirements set out in Sections 1 and 8 shall be understood to mean that the device, when used under the conditions and for the purposes intended, does not present a risk at all or presents a risk that is no more than the maximum acceptable risk related to the product's use which is consistent with a high level of protection for the safety and health of persons.				
10.	Chemical, physical, and biological properties				

10.1	Devices shall be designed and manufactured in such a way as to ensure that the characteristics and performance requirements referred to in Chapter I are fulfilled. Particular attention shall be paid to:				
	a. The choice of materials and substances used, particularly as regards toxicity and, where relevant, flammability;				
	b. The compatibility between the materials and substances used and biological tissues, cells, and body fluids, taking account of the intended purpose of the device and, where relevant, absorption, distribution, metabolism, and excretion;				
	c. The compatibility between the different parts of a device which consists of more than one implantable part;				
	d. The impact of processes on material properties;				
	e. Where appropriate, the results of biophysical or modeling research the validity of which has been demonstrated beforehand;				
	f. The mechanical properties of the materials used, reflecting, where appropriate, matters such as strength, ductility, fracture resistance, wear resistance and fatigue resistance;				
	g. Surface properties; and				
	h. The confirmation that the device meets any defined chemical and/or physical specifications.				
10.2	Devices shall be designed, manufactured, and packaged in such a way as to minimize the risk posed by contaminants and residues to patients, taking account of the intended purpose of the device, and to the persons involved in the transport, storage, and use of the devices. Particular attention shall be paid to tissues exposed to those contaminants and residues and to the duration and frequency of exposure.				
10.3	Devices shall be designed and manufactured in such a way that they can be used safely with the materials and substances, including gases, with which they enter into contact during their intended use; if the devices are intended to administer medicinal products they shall be designed and manufactured in such a way as to be compatible with the medicinal products concerned in accordance with the provisions and restrictions governing those medicinal products and that the performance of both the medicinal products and of the devices is maintained in accordance with their respective indications and intended use.				
10.4	Substances				
10.4.1	Design and manufacture of devices				
	Devices shall be designed and manufactured in such a way as to reduce as far as possible the risks posed by substances or particles, including wear debris, degradation products and processing residues that may be released from the device.				

	Devices, or those parts thereof or those materials used therein that: □ Are invasive and come into direct contact with the human body. □ (re)administer medicines, body liquids or other substances, including gases, to/from the body, or □ Transport or store such medicines, body fluids or substances, including gases, to be (re)administered to the body. shall only contain the following substances in a concentration that is above 0.1% weight by weight (w/w) where justified pursuant to Section 10.4.2.				
	a. Substances which are carcinogenic, mutagenic or toxic to reproduction ('CMR'), of category 1A or 1B, in accordance with Part 3 of Annex VI to Regulation (EC) No 1272/2008 of the European Parliament and of the Council (1), or				
	b. Substances having endocrine-disrupting properties for which there is scientific evidence of probable serious effects to human health and which are identified either in accordance with the procedure set out in Article 59 of Regulation (EC) No 1907/2006 of the European Parliament and of the Council (2) or, once a delegated act has been adopted by the Commission pursuant to the first subparagraph of Article 5(3) of Regulation (EU) No 528/2012 of the European Parliament and the Council (3), in accordance with the criteria that are relevant to human health amongst the criteria established therein.				
10.4.2	Justification regarding the presence of CMR and/or endocrine-disrupting substances				
	The justification for the presence of such substances shall be based upon: a. An analysis and estimation of potential patient or user exposure to the substance;				
	b. An analysis of possible alternative substances, materials or designs, including, where available, information about independent research, peer-reviewed studies, scientific opinions from relevant scientific committees and an analysis of the availability of such alternatives;				
	c. Argumentation as to why possible substance and/ or material substitutes, if available, or design changes, if feasible, are inappropriate in relation to maintaining the functionality, performance, and the benefit-risk ratios of the product; including taking into account if the intended use of such devices includes treatment of children or treatment of pregnant or breastfeeding women or treatment of other patient groups considered particularly vulnerable to such substances and/or materials; and				
	d. Where applicable and available, the latest relevant scientific committee guidelines in accordance with Sections 10.4.3 and 10.4.4.				
10.4.3	Guidelines on phthalates				

	For the purposes of Section 10.4., the Commission shall, as soon as possible and by 26 May 2018, provide the relevant scientific committee with a mandate to prepare guidelines that shall be ready before 26 May 2020. The mandate for the committee shall encompass at least a benefit-risk assessment of the presence of phthalates which belong to either of the groups of substances referred to in points (a) and (b) of Section 10.4.1. The benefit-risk assessment shall take into account the intended purpose and context of the use of the device, as well as any available alternative substances and alternative materials, designs or medical treatments. When deemed appropriate on the basis of the latest scientific evidence, but at least every five years, the guidelines shall be updated.				
10.4.4	Guidelines on other CMR and endocrine-disrupting substances				
	Subsequently, the Commission shall mandate the relevant scientific committee to prepare guidelines as referred to in Section 10.4.3. Also, for other substances referred to in points (a) and (b) of Section 10.4.1., where appropriate.				
10.4.5	Labeling				
	Where devices, parts thereof or materials used therein as referred to in Section 10.4.1. Contain substances referred to in points (a) or (b) of Section 10.4.1. in a concentration above 0.1% weight by weight (w/w), the presence of those substances shall be labeled on the device itself and/or on the packaging for each unit or, where appropriate, on the sales packaging, with the list of such substances. If the intended use of such devices includes treatment of children or treatment of pregnant or breastfeeding women or treatment of other patient groups considered particularly vulnerable to such substances and/or materials, information on residual risks for those patient groups and, if applicable, on appropriate precautionary measures shall be given in the instructions for use.				
10.5	Devices shall be designed and manufactured in such a way as to reduce as far as possible the risks posed by the unintentional ingress of substances into the device taking into account the device and the nature of the environment in which it is intended to be used.				
10.6	Devices shall be designed and manufactured in such a way as to reduce as far as possible the risks linked to the size and the properties of particles which are or can be released into the patient's or user's body, unless they come into contact with intact skin only. Special attention shall be given to nanomaterials.				
11.	Infection and microbial contamination				

11.1	Devices and their manufacturing processes shall be designed in such a way as to eliminate or to reduce as far as possible the risk of infection to patients, users, and, where applicable, other persons. The design shall: a. Reduce as far as possible and appropriate the risks from unintended cuts and pricks, such as needle stick injuries. b. Allow easy and safe handling. c. Reduce as far as possible any microbial leakage from the device and/or microbial exposure during use, and d. Prevent microbial contamination of the device or its content such as specimens or fluids.				
11.2	Where necessary devices shall be designed to facilitate their safe cleaning, disinfection, and/or re-sterilization.				
11.3	Devices labeled as having a specific microbial state shall be designed, manufactured, and packaged to ensure that they remain in that state when placed on the market and remain so under the transport and storage conditions specified by the manufacturer.				
11.4	Devices delivered in a sterile state shall be designed, manufactured, and packaged in accordance with appropriate procedures, to ensure that they are sterile when placed on the market and that, unless the packaging which is intended to maintain their sterile condition is damaged, they remain sterile, under the transport and storage conditions specified by the manufacturer, until that packaging is opened at the point of use. It shall be ensured that the integrity of that packaging is clearly evident to the final user.				
11.5	Devices labeled as sterile shall be processed, manufactured, packaged, and, sterilized by means of appropriate, validated methods.				
11.6	Devices intended to be sterilized shall be manufactured and packaged in appropriate and controlled conditions and facilities.				
11.7	Packaging systems for non-sterile devices shall maintain the integrity and cleanliness of the product and, where the devices are to be sterilized prior to use, minimize the risk of microbial contamination; the packaging system shall be suitable taking account of the method of sterilization indicated by the manufacturer.				
11.8	The labeling of the device shall distinguish between identical or similar devices placed on the market in both a sterile and a non-sterile condition additional to the symbol used to indicate that devices are sterile.				
12	Devices incorporating a substance considered to be a medicinal product and devices that are composed of substances or of combinations of substances that are absorbed by or locally dispersed in the human body.				
12.1	In the case of devices referred to in the first subparagraph of Article 1(8), the quality, safety, and usefulness of the substance which, if used separately, would be considered to be a medicinal product within the meaning of point (2) of Article 1 of Directive 2001/83/EC, shall be verified by analogy with the methods specified in Annex I to Directive 2001/83/EC, as required by the applicable conformity assessment procedure under this Regulation.				

12.2	Devices that are composed of substances or of combinations of substances that are intended to be introduced into the human body, and that are absorbed by or locally dispersed in the human body shall comply, where applicable and in a manner limited to the aspects not covered by this Regulation, with the relevant requirements laid down in Annex I to Directive 2001/83/EC for the evaluation of absorption, distribution, metabolism, excretion, local tolerance, toxicity, interaction with other devices, medicinal products or other substances and potential for adverse reactions, as required by the applicable conformity assessment procedure under this Regulation.				
13	Devices incorporating materials of biological origin				
13.1	For devices manufactured utilizing derivatives of tissues or cells of human origin which are non-viable or are rendered non-viable covered by this Regulation in accordance with point (g) of Article 1(6), the following shall apply: a. Donation, procurement, and testing of the tissues and cells shall be done in accordance with Directive 2004/23/EC; b. Processing, preservation, and any other handling of those tissues and cells or their derivatives shall be carried out so as to provide safety for patients, users, and, where applicable, other persons. In particular, safety with regard to viruses and other transmissible agents shall be addressed by appropriate methods of sourcing and by implementation of validated methods of elimination or inactivation in the course of the manufacturing process; c. The traceability system for those devices shall be complementary and compatible with the traceability and data protection requirements laid down in Directive 2004/23/EC and in Directive 2002/98/EC.				
13.2	For devices manufactured utilizing tissues or cells of animal origin, or their derivatives, which are non-viable or rendered non-viable the following shall apply: a. Where feasible taking into account the animal species, tissues, and cells of animal origin, or their derivatives, shall originate from animals that have been subjected to veterinary controls that are adapted to the intended use of the tissues. Information on the geographical origin of the animals shall be retained by manufacturers;				
13.3	13.3. For devices manufactured utilizing non-viable biological substances other than those referred to in Sections 13.1 and 13.2, the processing, preservation, testing, and handling of those substances shall be carried out so as to provide safety for patients, users, and, where applicable, other persons, including in the waste disposal chain. In particular, safety with regard to viruses and other transmissible agents shall be addressed by appropriate methods of sourcing and by implementation of validated methods of elimination or inactivation in the course of the manufacturing process.				

	b. Sourcing, processing, preservation, testing, and handling of tissues, cells, and substances of animal origin, or their derivatives, shall be carried out so as to provide safety for patients, users, and, where applicable, other persons. In particular safety with regard to viruses and other transmissible agents shall be addressed by implementation of validated methods of elimination or viral inactivation in the course of the manufacturing process, except when the use of such methods would lead to unacceptable degradation compromising the clinical benefit of the device;				
	c. In the case of devices manufactured utilizing tissues or cells of animal origin, or their derivatives, as referred to in Regulation (EU) No 722/2012 the particular requirements laid down in that Regulation shall apply.				
14	Construction of devices and interaction with their environment				
14.1	14.1. If the device is intended for use in combination with other devices or equipment the whole combination, including the connection system shall be safe and shall not impair the specified performance of the devices. Any restrictions on use applying to such combinations shall be indicated on the label and/or in the instructions for use. Connections which the user has to handle, such as fluid, gas transfer, electrical or mechanical coupling, shall be designed, and constructed in such a way as to minimize all possible risks, such as misconnection.				
14.2	14.2. Devices shall be designed and manufactured in such a way as to remove or reduce as far as possible: a. The risk of injury, in connection with their physical features, including the volume/pressure ratio, dimensional, and where appropriate ergonomic features;				
	b. Risks connected with reasonably foreseeable external influences or environmental conditions, such as magnetic fields, external electrical and electromagnetic effects, electrostatic discharge, radiation associated with diagnostic or therapeutic procedures, pressure, humidity, temperature, variations in pressure and acceleration or radio signal interferences;				
	c. The risks associated with the use of the device when it comes into contact with materials, liquids, and substances, including gases, to which it is exposed during normal conditions of use;				
	d. The risks associated with the possible negative interaction between software and the IT environment within which it operates and interacts;				
	e. The risks of accidental ingress of substances into the device;				
	f. The risks of reciprocal interference with other devices normally used in the investigations or for the treatment given; and				
	g. Risks arising where maintenance or calibration are not possible (as with implants), from aging of materials used or loss of accuracy of any measuring or control mechanism.				
14.3	Devices shall be designed and manufactured in such a way as to minimize the risks of fire or explosion during normal use and in single fault condition. Particular attention shall be paid to devices the intended use of which includes exposure to or use in association with flammable or explosive substances or substances which could cause combustion.				

14.4	Devices shall be designed and manufactured in such a way that adjustment, calibration, and maintenance can be done safely and effectively.				
14.5	Devices that are intended to be operated together with other devices or products shall be designed and manufactured in such a way that the interoperability and compatibility are reliable and safe.				
14.6	Any measurement, monitoring, or display scale shall be designed and manufactured in line with ergonomic principles, taking account of the intended purpose, users, and the environmental conditions in which the devices are intended to be used.				
14.7	Devices shall be designed and manufactured in such a way as to facilitate their safe disposal and the safe disposal of related waste substances by the user, patient, or other person. To that end, manufacturers shall identify and test procedures and measures as a result of which their devices can be safely disposed after use. Such procedures shall be described in the instructions for use.				
15	Devices with a diagnostic or measuring function				
15.1	Diagnostic devices and devices with a measuring function, shall be designed and manufactured in such a way as to provide sufficient accuracy, precision, and stability for their intended purpose, based on appropriate scientific and technical methods. The limits of accuracy shall be indicated by the manufacturer.				
15.2	The measurements made by devices with a measuring function shall be expressed in legal units conforming to the provisions of Council Directive 80/181/EEC (1).				
16	Protection against radiation				
16.1.	General:				
	a. Devices shall be designed, manufactured, and packaged in such a way that exposure of patients, users, and other persons to radiation is reduced as far as possible, and in a manner that is compatible with the intended purpose, whilst not restricting the application of appropriate specified levels for therapeutic and diagnostic purposes.				
	b. The operating instructions for devices emitting hazardous or potentially hazardous radiation shall contain detailed information as to the nature of the emitted radiation, the means of protecting the patient and the user, and on ways of avoiding misuse and of reducing the risks inherent to installation as far as possible and appropriate. Information regarding the acceptance and performance testing, the acceptance criteria, and the maintenance procedure shall also be specified.				
16.2	Intended radiation				
	a. Where devices are designed to emit hazardous, or potentially hazardous, levels of ionizing and/ or non-ionizing radiation necessary for a specific medical purpose the benefit of which is considered to outweigh the risks inherent to the emission, it shall be possible for the user to control the emissions. Such devices shall be designed and manufactured to ensure reproducibility of relevant variable parameters within an acceptable tolerance.				

	b. Where devices are intended to emit hazardous, or potentially hazardous, ionizing, and/or non-ionizing radiation, they shall be fitted, where possible, with visual displays and/or audible warnings of such emissions.				
16.3	Devices shall be designed and manufactured in such a way that exposure of patients, users, and other persons to the emission of unintended, stray or scattered radiation is reduced as far as possible. Where possible and appropriate, methods shall be selected which reduce the exposure to radiation of patients, users, and other persons who may be affected.				
16.4	Ionizing radiation				
	a. Devices intended to emit ionizing radiation shall be designed and manufactured taking into account the requirements of the Directive 2013/59/Euratom laying down basic safety standards for protection against the dangers arising from exposure to ionizing radiation.				
	b. Devices intended to emit ionizing radiation shall be designed and manufactured in such a way as to ensure that, where possible, taking into account the intended use, the quantity, geometry, and quality of the radiation emitted can be varied and controlled, and, if possible, monitored during treatment.				
	c. Devices emitting ionizing radiation intended for diagnostic radiology shall be designed and manufactured in such a way as to achieve an image and/or output quality that are appropriate to the intended medical purpose whilst minimizing radiation exposure of the patient and user.				
	d. Devices that emit ionizing radiation and are intended for therapeutic radiology shall be designed and manufactured in such a way as to enable reliable monitoring and control of the delivered dose, the beam type, energy, and, where appropriate, the quality of radiation.				
17	Electronic programmable systems—devices that incorporate electronic programmable systems and software that are devices in themselves				
17.1	Devices that incorporate electronic programmable systems, including software, or software that are devices in themselves, shall be designed to ensure repeatability, reliability, and performance in line with their intended use. In the event of a single fault condition, appropriate means shall be adopted to eliminate or reduce as far as possible consequent risks or impairment of performance.				
17.2	For devices that incorporate software or for software that are devices in themselves, the software shall be developed and manufactured in accordance with the state of the art taking into account the principles of development life cycle, risk management, including information security, verification, and validation.				
17.3	Software referred to in this Section that is intended to be used in combination with mobile computing platforms shall be designed and manufactured taking into account the specific features of the mobile platform (e.g., size, and contrast ratio of the screen) and the external factors related to their use (varying environment as regards level of light or noise).				
17.4	Manufacturers shall set out minimum requirements concerning hardware; IT networks characteristics and IT security measures, including protection against unauthorized access, necessary to run the software as intended.				

18	Active devices and devices connected to them				
	18.1. For non-implantable active devices, in the event of a single fault condition, appropriate means shall be adopted to eliminate or reduce as far as possible consequent risks.				
	18.2. Devices where the safety of the patient depends on an internal power supply shall be equipped with a means of determining the state of the power supply and an appropriate warning or indication for when the capacity of the power supply becomes critical. If necessary, such warning or indication shall be given prior to the power supply becoming critical.				
	18.3. Devices where the safety of the patient depends on an external power supply shall include an alarm system to signal any power failure.				
	18.4. Devices intended to monitor one or more clinical parameters of a patient shall be equipped with appropriate alarm systems to alert the user of situations which could lead to death or severe deterioration of the patient's state of health.				
	18.5. Devices shall be designed and manufactured in such a way as to reduce as far as possible the risks of creating electromagnetic interference which could impair the operation of the device in question or other devices or equipment in the intended environment.				
	18.6. Devices shall be designed and manufactured in such a way as to provide a level of intrinsic immunity to electromagnetic interference such that is adequate to enable them to operate as intended.				
	18.7. Devices shall be designed and manufactured in such a way as to avoid, as far as possible, the risk of accidental electric shocks to the patient, user or any other person, both during normal use of the device and in the event of a single fault condition in the device, provided the device is installed and maintained as indicated by the manufacturer.				
	18.8. Devices shall be designed and manufactured in such a way as to protect, as far as possible, against unauthorized access that could hamper the device from functioning as intended.				
19	Particular requirements for active implantable devices				
	19.1. Active implantable devices shall be designed and manufactured in such a way as to remove or minimize as far as possible: a. Risks connected with the use of energy sources with particular reference, where electricity is used, to insulation, leakage currents and overheating of the devices. b. Risks connected with medical treatment, in particular those resulting from the use of defibrillators or high-frequency surgical equipment, and c. Risks which may arise where maintenance and calibration are impossible, including: □ Excessive increase of leakage currents. □ Aging of the materials used. □ Excess heat generated by the device. □ Decreased accuracy of any measuring or control mechanism.				

	19.2. Active implantable devices shall be designed and manufactured in such a way as to ensure □ If applicable, the compatibility of the devices with the substances they are intended to administer, and □ The reliability of the source of energy.				
	19.3. Active implantable devices and, if appropriate, their component parts shall be identifiable to allow any necessary measure to be taken following the discovery of a potential risk in connection with the devices or their component parts.				
	19.4. Active implantable devices shall bear a code by which they and their manufacturer can be unequivocally identified (particularly with regard to the type of device and its year of manufacture); it shall be possible to read this code, if necessary, without the need for a surgical operation.				
20	Protection against mechanical and thermal risks				
	20.1. Devices shall be designed and manufactured in such a way as to protect patients and users against mechanical risks connected with, for example, resistance to movement, instability, and moving parts.				
	20.2. Devices shall be designed and manufactured in such a way as to reduce to the lowest possible level the risks arising from vibration generated by the devices, taking account of technical progress and of the means available for limiting vibrations, particularly at source, unless the vibrations are part of the specified performance.				
	20.3. Devices shall be designed and manufactured in such a way as to reduce to the lowest possible level the risks arising from the noise emitted, taking account of technical progress and of the means available to reduce noise, particularly at source, unless the noise emitted is part of the specified performance.				
	20.4. Terminals and connectors to the electricity, gas or hydraulic and pneumatic energy supplies which the user or other person has to handle, shall be designed and constructed in such a way as to minimize all possible risks.				
	20.5. Errors likely to be made when fitting or refitting certain parts which could be a source of risk shall be made impossible by the design and construction of such parts or, failing this, by information given on the parts themselves and/or their housings. The same information shall be given on moving parts and/or their housings where the direction of movement needs to be known in order to avoid a risk.				
	20.6. Accessible parts of devices (excluding the parts or areas intended to supply heat or reach given temperatures) and their surroundings shall not attain potentially dangerous temperatures under normal conditions of use.				
21.	Protection against the risks posed to the patient or user by devices supplying energy or substances				
21.1	21.1. Devices for supplying the patient with energy or substances shall be designed and constructed in such a way that the amount to be delivered can be set and maintained accurately enough to ensure the safety of the patient and of the user.				

	21.2. Devices shall be fitted with the means of preventing and/or indicating any inadequacies in the amount of energy delivered or substances delivered which could pose a danger. Devices shall incorporate suitable means to prevent, as far as possible, the accidental release of dangerous levels of energy or substances from an energy and/or substance source.				
	21.3. The function of the controls and indicators shall be clearly specified on the devices. Where a device bears instruction required for its operation or indicates operating or adjustment parameters by means of a visual system, such information shall be understandable to the user and, as appropriate, the patient.				
22.	Protection against the risks posed by medical devices intended by the manufacturer for use by lay persons				
22.1	Devices for use by lay persons shall be designed and manufactured in such a way that they perform appropriately for their intended purpose taking into account the skills and the means available to lay persons and the influence resulting from variation that can be reasonably anticipated in the lay person's technique and environment. The information and instructions provided by the manufacturer shall be easy for the lay person to understand and apply.				
	22.2. Devices for use by lay persons shall be designed and manufactured in such a way as to: □ Ensure that the device can be used safely and accurately by the intended user at all stages of the procedure, if necessary after appropriate training and/or information. □ Reduce, as far as possible and appropriate, the risk from unintended cuts and pricks such as needle stick injuries, and □ Reduce as far as possible the risk of error by the intended user in the handling of the device and, if applicable, in the interpretation of the results.				
22.3	Devices for use by lay persons shall, where appropriate, include a procedure by which the lay person: □ Can verify that, at the time of use, the device will perform as intended by the manufacturer, and □ If applicable, is warned if the device has failed to provide a valid result.				
	Requirements regarding the information supplied with the device				
23.	Label and instructions for use				
	23.1. General requirements regarding the information supplied by the manufacturer. Each device shall be accompanied by the information needed to identify the device and its manufacturer, and by any safety and performance information relevant to the user, or any other person, as appropriate. Such information may appear on the device itself, on the packaging or in the instructions for use, and shall, if the manufacturer has a website, be made available and kept up to date on the website, taking into account the following:				

	a. The medium, format, content, legibility, and location of the label and instructions for use shall be appropriate to the particular device, its intended purpose and the technical knowledge, experience, education or training of the intended user(s). In particular, instructions for use shall be written in terms readily understood by the intended user and, where appropriate, supplemented with drawings and diagrams.				
	b. The information required on the label shall be provided on the device itself. If this is not practicable or appropriate, some or all of the information may appear on the packaging for each unit, and/or on the packaging of multiple devices.				
	c. Labels shall be provided in a human-readable format and may be supplemented by machine-readable information, such as radio-frequency identification ('RFID') or bar codes.				
	d. Instructions for use shall be provided together with devices. By way of exception, instructions for use shall not be required for class I and class IIa devices if such devices can be used safely without any such instructions and unless otherwise provided for elsewhere in this Section.				
	e. Where multiple devices are supplied to a single user and/or location, a single copy of the instructions for use may be provided if so agreed by the purchaser who in any case may request further copies to be provided free of charge.				
	f. Instructions for use may be provided to the user in non-paper format (e.g., electronic) to the extent, and only under the conditions, set out in Regulation (EU) No 207/2012 or in any subsequent implementing rules adopted pursuant to this Regulation.				
	g. Residual risks which are required to be communicated to the user and/or other person shall be included as limitations, contra-indications, precautions, or warnings in the information supplied by the manufacturer.				
	h. Where appropriate, the information supplied by the manufacturer shall take the form of internationally recognized symbols. Any symbol or identification color used shall conform to the harmonized standards or CS. In areas for which no harmonized standards or CS exist, the symbols and colors shall be described in the documentation supplied with the device.				
	23.2. Information on the label The label shall bear all of the following particulars: a. The name or trade name of the device;				
	b. The details strictly necessary for a user to identify the device, the contents of the packaging and, where it is not obvious for the user, the intended purpose of the device;				
	c. The name, registered trade name or registered trade mark of the manufacturer and the address of its registered place of business;				
	d. If the manufacturer has its registered place of business outside the Union, the name of the authorized representative and address of the registered place of business of the authorized representative;				

	e. Where applicable, an indication that the device contains or incorporates: □ A medicinal substance, including a human blood or plasma derivative, or □ Tissues or cells, or their derivatives, of human origin, or □ Tissues or cells of animal origin, or their derivatives, as referred to in Regulation (EU) No 722/2012;				
	f. Where applicable, information labeled in accordance with Section 10.4.5.				
	g. The lot number or the serial number of the device preceded by the words LOT NUMBER or SERIAL NUMBER or an equivalent symbol, as appropriate;				
	h. The UDI carrier referred to in Article 27(4) and Part C of Annex VII;				
	i. An unambiguous indication of t the time limit for using or implanting the device safely, expressed at least in terms of year and month, where this is relevant;				
	j. Where there is no indication of the date until when it may be used safely, the date of manufacture. This date of manufacture may be included as part of the lot number or serial number, provided the date is clearly identifiable;				
	k. An indication of any special storage and/or handling condition that applies;				
	l. If the device is supplied sterile, an indication of its sterile state and the sterilization method;				
	m. Warnings or precautions to be taken that need to be brought to the immediate attention of the user of the device, and to any other person. This information may be kept to a minimum in which case more detailed information shall appear in the instructions for use, taking into account the intended users;				
	n. If the device is intended for single use, an indication of that fact. A manufacturer's indication of single use shall be consistent across the Union;				
	o. If the device is a single-use device that has been reprocessed, an indication of that fact, the number of reprocessing cycles already performed, and any limitation as regards the number of reprocessing cycles;				
	p. If the device is custom-made, the words 'custom-made device';				
	q. An indication that the device is a medical device. If the device is intended for clinical investigation only, the words 'exclusively for clinical investigation';				
	r. In the case of devices that are composed of substances or of combinations of substances that are intended to be introduced into the human body via a body orifice or applied to the skin and that are absorbed by or locally dispersed in the human body, the overall qualitative composition of the device and quantitative information on the main constituent or constituents responsible for achieving the principal intended action;				
	s. For active implantable devices, the serial number, and for other implantable devices, the serial number or the lot number.				
23.3	Information on the packaging which maintains the sterile condition of a device ('sterile packaging')				

	The following particulars shall appear on the sterile packaging: a. An indication permitting the sterile packaging to be recognized as such. b. A declaration that the device is in a sterile condition. c. The method of sterilization. d. The name and address of the manufacturer. e. A description of the device. f. If the device is intended for clinical investigations, the words 'exclusively for clinical investigations,' g. If the device is custom-made, the words 'custom-made device,' h. The month and year of manufacture. i. An unambiguous indication of the time limit for using or implanting the device safely expressed at least in terms of year and month, and j. An instruction to check the instructions for use for what to do if the sterile packaging is damaged or unintentionally opened before use.				
	23.4. Information in the instructions for use				
	The instructions for use shall contain all of the following particulars: a. The particulars referred to in points (a), (c), (e), (f), (k), (l), (n) and (r) of Section 23.2;				
	b. The device's intended purpose with a clear specification of indications, contra-indications, the patient target group or groups, and of the intended users, as appropriate;				
	c. Where applicable, a specification of the clinical benefits to be expected.				
	d. Where applicable, links to the summary of safety and clinical performance referred to in Article 32;				
	(e) the performance characteristics of the device;				
	f. Where applicable, information allowing the healthcare professional to verify if the device is suitable and select the corresponding software and accessories;				
	g. Any residual risks, contra-indications, and any undesirable side-effects, including information to be conveyed to the patient in this regard;				
	h. Specifications the user requires to use the device appropriately, e.g., if the device has a measuring function, the degree of accuracy claimed for it;				
	i. Details of any preparatory treatment or handling of the device before it is ready for use or during its use, such as sterilization, final assembly, calibration, etc., including the levels of disinfection required to ensure patient safety and all available methods for achieving those levels of disinfection;				
	j. Any requirements for special facilities, or special training, or particular qualifications of the device user and/or other persons;				

	k. The information needed to verify whether the device is properly installed and is ready to perform safely and as intended by the manufacturer, together with, where relevant: □ Details of the nature, and frequency, of preventive and regular maintenance, and of any preparatory cleaning or disinfection. □ Identification of any consumable components and how to replace them. □ Information on any necessary calibration to ensure that the device operates properly and safely during its intended lifetime, and □ Methods for eliminating the risks encountered by persons involved in installing, calibrating or servicing devices;				
	l. If the device is supplied sterile, instructions in the event of the sterile packaging being damaged or unintentionally opened before use;				
	m. If the device is supplied non-sterile with the intention that it is sterilized before use, the appropriate instructions for sterilization;				
	n. If the device is reusable, information on the appropriate processes for allowing reuse, including cleaning, disinfection, packaging, and, where appropriate, the validated method of re-sterilization appropriate to the Member State or Member States in which the device has been placed on the market. Information shall be provided to identify when the device should no longer be reused, e.g., signs of material degradation or the maximum number of allowable reuses;				
	o. An indication, if appropriate, that a device can be reused only if it is reconditioned under the responsibility of the manufacturer to comply with the general safety and performance requirements;				
	p. If the device bears an indication that it is for single use, information on known characteristics and technical factors known to the manufacturer that could pose a risk if the device were to be re-used. This information shall be based on a specific section of the manufacturer's risk management documentation, where such characteristics and technical factors shall be addressed in detail. If in accordance with point (d) of Section 23.1. no instructions for use are required, this information shall be made available to the user upon request;				
	q. For devices intended for use together with other devices and/or general-purpose equipment: □ Information to identify such devices or equipment, in order to obtain a safe combination, and/or □ Information on any known restrictions to combinations of devices and equipment;				
	r. If the device emits radiation for medical purposes: □ Detailed information as to the nature, type, and where appropriate, the intensity and distribution of the emitted radiation. □ The means of protecting the patient, user, or other person from unintended radiation during use of the device;				

	s. Information that allows the user and/or patient to be informed of any warnings, precautions, contra-indications, measures to be taken and limitations of use regarding the device. That information shall, where relevant, allow the user to brief the patient about any warnings, precautions, contra-indications, measures to be taken and limitations of use regarding the device. The information shall cover, where appropriate: □ Warnings, precautions, and/or measures to be taken in the event of malfunction of the device or changes in its performance that may affect safety. □ Warnings, precautions, and/or measures to be taken as regards the exposure to reasonably foreseeable external influences or environmental conditions, such as magnetic fields, external electrical and electromagnetic effects, electrostatic discharge, radiation associated with diagnostic or therapeutic procedures, pressure, humidity, or temperature. □ Warnings, precautions, and/or measures to be taken as regards the risks of interference posed by the reasonably foreseeable presence of the device during specific diagnostic investigations, evaluations, or therapeutic treatment or other procedures such as electromagnetic interference emitted by the device affecting other equipment. —if the device is intended to administer medicinal products, tissues or cells of human or animal origin, or their derivatives, or biological substances, any limitations or incompatibility in the choice of substances to be delivered. —warnings, precautions, and/or limitations related to the medicinal substance or biological material that is incorporated into the device as an integral part of the device; and —precautions related to materials incorporated into the device that contain or consist of CMR substances or endocrine-disrupting substances, or that could result in sensitization or an allergic reaction by the patient or user;				
	t. In the case of devices that are composed of substances or of combinations of substances that are intended to be introduced into the human body and that are absorbed by or locally dispersed in the human body, warnings, and precautions, where appropriate, related to the general profile of interaction of the device and its products of metabolism with other devices, medicinal products and other substances as well as contra-indications, undesirable side-effects and risks relating to overdose;				
	u. In the case of implantable devices, the overall qualitative and quantitative information on the materials and substances to which patients can be exposed;				
	v. Warnings or precautions to be taken in order to facilitate the safe disposal of the device, its accessories, and the consumables used with it, if any. This information shall cover, where appropriate:—infection or microbial hazards such as explants, needles, or surgical equipment contaminated with potentially infectious substances of human origin, and—physical hazards such as from sharps. If in accordance with the point (d) of Section 23.1 no instructions for use are required, this information shall be made available to the user upon request;				
	w. For devices intended for use by lay persons, the circumstances in which the user should consult a healthcare professional;				

	(x) for the devices covered by this Regulation pursuant to Article 1(2), information regarding the absence of a clinical benefit and the risks related to use of the device;				
	y. Date of issue of the instructions for use or, if they have been revised, date of issue and identifier of the latest revision of the instructions for use;				
	z. A notice to the user and/or patient that any serious incident that has occurred in relation to the device should be reported to the manufacturer and the competent authority of the Member State in which the user and/or patient is established;				
	aa. Information to be supplied to the patient with an implanted device in accordance with Article 18;				
	ab. For devices that incorporate electronic programmable systems, including software, or software that are devices in themselves, minimum requirements concerning hardware, IT networks characteristics and IT security measures, including protection against unauthorized access, necessary to run the software as intended.				

Simple easy to use templates and guides for completing a General Safety and Performance Requirements checklist can be downloaded from https://scientistssanctuary.com/

References

1. MHRA (Medicines and Healthcare products Regulatory Agency) Medical Device Alert: DePuy ASR acetabular cups used in hip resurfacing arthroplasty and total hip replacement. MDA/2010/044. 25 May 2010. 2010a.
2. American Medical Association 2003 World Medical Association Declaration of Helsinki. Ethical Principles for Medical Research Involving Human Subjects. Clinical Review and Education. 2191 – 2194.
3. World Health Organization 2001 Bulletin of the World Health Organization. World Medical Association Declaration of Helsinki. Ethical Principles for Medical Research Involving Human Subjects.
4. Carlson R V., K. M. Boyd and D. J. Webb. 2004. The revision of the Declaration of Helsinki: past, present and future. British Journal of Clinical Pharmacology. 57: 695 – 713.
5. Report of the Interregional Meeting on Maintenance and Repair of Health Care Equipment, Nicosia, Cyprus, November 1996. WHO/ SHS/NHP/87.5, WHO, Geneva, 1987.
6. Prage, L.: Guidelines for Support to Procurement, Operation and Maintenance of Scientific Equipment in Developing Countries. IFS/

SAREC, Stockholm, Sweden 1987

7. Policy on Acquiring Medical Equipment Through Quotation/Tender and Donation (Loose Minute). Ministry of Health, Nairobi, Kenya, 1987.
8. Study on Drugs, Equipment and Supplies and the Role of Christian Medical Association of India. CMAI, New Delhi, 1987.
9. International Health Relief Assistance: A Guide for Effective Aid. Pan American Health Organization/WHO Regional office for the Americas, Washington, 1990.
10. Guidelines for Donors and Recipients of Equipment Donations. Joint Medical Stores, Kampala, Uganda, 1990.
11. Guidelines for Donors and Recipients of Pharmaceutical Donations. Second Revised Edition. Christian Medical Commission, World Council of Churches, Geneva, 1990.
12. Bloom, G., Temple-Bird, C.: Medical Equipment in Sub-Saharan Africa: A Framework for Policy-Formulation. IDS Research Report No. 19, WHO/SHS/NHP/90.6, WHO, Geneva, 1990.
13. Guidelines for Purchasing of Medical Equipment. Presbyterian Technical Services, Presbyterian Church in Cameroon, Kumba, Cameroon, 1991.
14. Kachienga, M.O.: An Overview: Procurement of Health Care Equipment for Public Health Facilities. Medicus, pp. 20-23, 1992.
15. IMDG, Donating and Selling Used Medical Equipment. Health Devices, ECRI, Plymouth Meeting, PA 21(9) pp.295-7, 1992.
16. Issakov, A.: Service and Maintenance in Developing Countries, pp. 21-28 in: Medical Devices: International Perspectives on Health and Safety. Ed. Van Gruting C.W.D. Elsevier, Amsterdam, 1994.
17. Executive Report of the Regional Workshop on Health care Technology in Sub-Saharan Region, Somerset West, South Africa, April 1994. South African Medical Research Council, Tygerberg, 1994.
18. Falcitelli, G.: Biomedical Technology and Clinical Engineering in Cooperation Programmes with Developing Countries: Metholological Guidelines. p. 13 in: Proceedings of the Regional Workshop on Health Care Technology in the Sub-Saharan Region, Somerset West, South Africa. April 1994 South African Medical Research Council, Tygerberg, 1994.
19. Temple-Bird, C.: Methodology for the Analysis of a National Medical

Equipment Management System Based on a Case Study in Botswana p. 33, ibid.

20. Guidelines: Medical Equipment Donations. Technical Library, FAKT, in: CMC Publication CONTACT 10/94. CMC, World Council of Churches, Geneva, 1994.
21. Guidelines for Medical Equipment Donation. American College of Clinical Engineering, Plymouth Meeting, PA, 1995.
22. Madani, M.A.: Medical Equipment Retirement and Donation Criteria. Presentation to the 30th AAMI Annual Meeting, Anaheim, CA, USA, May 1995.
23. Guidelines on Donations of Drugs and Medical Equipment to the Health Sector for Tanzania Mainland. Ministry of Health, Dar-es-Salaam, Tanzania, 1995.
24. Guidelines for Drug Donations. WHO/DAP/96.2, WHO, Geneva, 1996.
25. Wang, B.: The Facts on Used Equipment Export. Biomedical Technology Management, p. 64, 1996.
26. A Disaster Relief Supply Management System (SUMA). Pan American Health Organization/WHO Regional Office for the Americas, Washington, 1996.
27. Application of risk management to medical devices (2007). International Organization for Standardization (ISO 14971:2007 Standard). https://www.iso.org/obp/ui/#iso:std:iso:14971:ed-2:v2:en (accessed on 17 March 2020).
28. Clinical investigations (2010). Global Harmonization Task Force (GHTF/SG5/ N3:2010). http://www.imdrf.org/docs/ghtf/final/sg5/technical-docs/ghtf-sg5-n3-clinical-investigations-100212.pdf (accessed on 17 March 2020).
29. Development of medical device policies (2011). Geneva: World Health Organization. WHO Medical Device Technical Series. http://apps.who.int/medicinedocs/documents/s21559en/s21559en.pdf (accessed on 17 March 2020).
30. EU Council Directive 93/42/EEC of 14 June (1993) concerning medical devices http://www.imdrf.org/docs/ghtf/archived/sg4/technical-docs/ghtf-sg4-guidelinesauditing-qms-part-1-general-requirements-080827.pdf (accessed on 17 March 2020).
31. Global Medical Devices Nomenclature System (GMDN) [website]

(https://www.gmdnagency.org).(accessed on 17 March 2020).

32. Good Governance for Medicines. Model Framework (2014). Geneva: World Health Organization. http://apps.who.int/iris/bitstream/10665/129495/1/9789241507516_eng.pdf?ua=1&ua=1 (accessed on 17 March 2020).
33. Guiding Principles for Regulatory Quality and Performance (2005). Paris: Organization for Economic Cooperation and Development (OECD). http://www.oecd.org/fr/reformereg/34976533.pdf (accessed on 17 March 2020).
34. Medical Device Regulations: Global Overview and Guiding Principles (2003). Geneva: World Health Organization. http://www.who.int/medical_devices/publications/en/MD_Regulations.pdf (accessed on 17 March 2020).
35. Regulation (EU) 2017/745 of the European Parliament and of the Council of 5 April 2017 on medical devices, amending Directive 2001/83/EC, Regulation (EC) No 178/2002 and Regulation (EC) No 1223/2009 and repealing Council Directives 90/385/EEC and 93/42/EEC
36. Principles of conformity assessment for medical devices (2012). Global Harmonization Task Force, (GHTF/SG1/N78:2012). http://www.imdrf.org/docs/ghtf/final/sg1/technical-docs/ghtf-sg1-n78-2012-conformity-assessment-medicaldevices-121102.pdf (accessed on 17 March 2020).
37. Principles of medical devices classification (2012). Global Harmonization Task Force; (GHTF/SG1/N77:2012; http://www.imdrf.org/docs/ghtf/final/sg1/technicaldocs/ghtf-sg1-n77-2012-principles-medical-devices-classification-121102.pdf (accessed on 17 March 2020).
38. Quality management system – Medical Devices – Guidance on corrective action and preventive action and related QMS processes (2010). Global Harmonization Task Force; (GHTF/SG3/N18:2010; http://www.imdrf.org/docs/ghtf/final/sg3/technical-docs/ghtf-sg3-n18-2010-qms-guidance-on-corrective-preventativeaction-101104.pdf (accessed on 17 March 2020).
39. Regulatory framework for control of refurbished medical devices (2012). Asia Pacific Economic Cooperation. http://mddb.apec.org/Documents/2012/MAG/WKSP1/12_mag_wksp1_011.pdf (accessed on 17 March 2020).

40. Regulatory harmonization and convergence. Silver Spring (MD): US Food and Drug Administration (US FDA) (http://www.fda.gov/BiologicsBloodVaccines/InternationalActivities/ucm271079.htm, (accessed on 17 March 2020).
41. Role of standards in the assessment of medical devices (2008). Global Harmonization Task Force; (GHTF/SG1/N044:2008). http://www.imdrf.org/docs/ghtf/final/sg1/procedural-docs/ghtf-sg1-n044-2008-standards-in-assessment-of-medicaldevices-080305.pdf (accessed on 17 March 2020).
42. Safety aspects – Guidelines for their inclusion in standards (2014). Geneva: International Organization for Standardization; (ISO/IEC Guide 51:2014. http://www.iso.org/iso/catalogue_detail.htm?csnumber=53940 (accessed on 17 March 2020).
43. WHO global governance http://www.who.int/healthpromotion/conferences/9gchp/good-governance/en/ (accessed on 17 March 2020).
44. WHO good regulatory practices: Guidelines for national regulatory authorities for medical products. Geneva: World Health Organization (under development, 2017).
45. Emergency Relief Items: Compendium of Basic Specifications, Volume 2: Medical Supplies and Equipment, UNDP/WHO, Geneva, 1996.
46. Rosenblatt, W.H.: Surplus Medical Supplies. To be published in World Health Magazine, 1997.
47. Chang, T.: Medical Device Donations to Developing Countries: Things to Consider. Poster Presentation at the 32nd AAMI Annual Meeting, Washington, D.C., USA, June 1997.
48. Heimann, P., Poluta, M.A.: Health Technology Management in the Sub-Saharan Region as a Prerequisite for Optimizing the Donor Aid Intervention Process. (In press) WHO, ARA, Geneva, 1997.
49. World Medical Association Declaration of Helsinki: Ethical Principles For Medical Research Involving Human Subjects. 2013. American Medical Association. JAMA, 310: 2191 – 2194.
50. Gmdnagency.com. Global Medical Device Nomenclature; User Guide. Available from: http://www.gmdnagency.com/ [last cited on 2010 Jan 10].
51. Gmdnagency.com. Global Medical Device Nomenclature. Available from: http://www.gmdnagency.com/?id=userinfo [last cited on 2010 Jan 14].

52. Ghtf.org. Global Harmonization Task Force; Generating the GMDN Process and Development. Available from: http//www.ghtf.org/meetings/ conferences/8thconference/.../Dillard.ppt [last cited on 2009 Nov 25].
53. Ahrq.gov. Agency for Health Research and quality; Brockton Hefflin; Center for Devices and Radiological Health; Food and Drug Administration. Available from: http//www.healthit.ahrq.gov/portal/server.pt?open=18. ppt [last cited on 2009 Dec 18].
54. Gmdnagency.com. Global Medical Device Nomenclature. Available from: http://www.gmdnagency.com/?id=userinfo [last cited on 2010 Feb 02].
55. Ahwp.info. Asian Harmonization Working Party; Report from AHWP STG Nomenclature. Available from: http://www.ahwp.info/ClientFolder/ AHWP/Library/Tree/4_Technical_Committee/3_TC_Minutes/The_9th_AHWP_Technical_Committee_Meeting/14_Medical_Device_ Nomenclautre.pdf [last cited on 2010 Feb 02].

INDEX

W